D0629259

CORPORATE SOCIAL RESPONSIBILITY FOR
A COMPETITIVE EDGE IN EMERGING MARKETS

CSR STRATEGIES

CORPORATE SOCIAL RESPONSIBILITY FOR A COMPETITIVE EDGE IN EMERGING MARKETS

SRI URIP

John Wiley & Sons (Asia) Pte. Ltd.

Published in 2010 by John Wiley & Sons (Asia) Pte. Ltd.

2 Clementi Loop, #02-01, Singapore 129809

Other Wiley Editorial Offices

John Wiley & Sons, 111 River Street, Hoboken, NJ 07030, USA
John Wiley & Sons, The Atrium, Southern Gate, Chichester, West Sussex, P019 8SQ, United Kingdom
John Wiley & Sons (Canada) Ltd., 5353 Dundas Street West, Suite 400, Toronto, Ontario, M9B 6HB, Canada
John Wiley & Sons Australia Ltd, 42 McDougall Street, Milton, Queensland 4064, Australia
Wiley-VCH, Boschstrasse 12, D-69469 Weinheim, Germany

Library of Congress Cataloging-in-Publication Data

ISBN: 978-0-47082520-4

Typeset in 10.5/14pt Plantin-Light by Thomson Digital

Printed in Singapore by Saik Wah Press Pte. Ltd.

10 9 8 7 6 5 4 3 2 1

Dedicated lovingly to my family;
my husband, Urip; my two children,
Tya and Tinggar; my son
and daughter-in-law,
Dwi Tjahyanto and Indira,
and my four grandchildren
Pandya, Chandrika,
Dayita, and Prajesa

Contents

Acknowledgments

CSR Strategies: Corporate Social Responsibility for a Competitive Edge in Emerging Markets originally started from material I prepared for LKDI (*Lembaga Komisaris dan Direktur Indonesia*—Indonesian Institute of Commissioners and Directors), APB Group, Ford Foundation Indonesia, Polling Centre, and other organizations. As corporate social responsibility (CSR) was becoming a more important, compulsory for many companies in Indonesia, I began to write a comprehensive draft based on my long experience in practicing this philosophy, which is now the first manuscript of this book.

I want to express my special gratitude and thanks to the late Ian Maxwell Harris and to Merina Purbo for kindly taking the time to help with revisions and editorial advice. Ian, my friend and big supporter, thank you for having understood my line of thoughts. You will be greatly missed. Also to Yonni Hadipradja, my nephew, who helped me with the preparation of all graphs and pictures. I really appreciate your effort in making sure that they were developed at the right stage for printing.

My great thanks goes to Maurits Lalisang, CEO of Unilever Indonesia, for his permission to publish the intelligence I gained through my 30 years of practicing CSR within Unilever, Indonesia. Thank you also for allowing me to share the continuing progress of all CSR activities after my retirement.

I also would like to thank Daniel Chandranayagam, Editor of *CSR Digest* for helping me in finalizing the book with the last revisions and editorial advice.

Of course, much of this book would not have been possible without the extraordinary support from several prominent organizations, whose application of CSR principles as an integral part of the business strategy have strongly positioned these companies as industry leaders continuously. I would like to express my appreciation and thanks to my colleagues in the following important organizations for their permission to publish in detail their company's CSR best practices and for all other refreshing perspectives given to me.

Josef Bataona, Human Resources/Corporate Social Responsibility Director, PT Unilever, Indonesia.

Manggi T. Habir, Chairman Board of Supervisors, Danamon Peduli Foundation.

Risa Bhinekawati, Chairperson/Executive Director, Danamon Peduli Foundation.

Jacobus Busono, Founder and President Director Pura Group.

Hasmo Soejono, Personal Secretary to the President Director Pura Group.

Angky Tisnadisastra, Director of PT Astra International, Tbk.

Arief Istanto, Senior Vice President, Chief Corporate Security; Environment & Social Responsibility, PT Astra International, Tbk.

Aminuddin, Chairman, Yayasan Dharma Bhakti Astra (YDBA), PT Astra International, Tbk.

Soepardi, General Manager, Yayasan Dharma Bhakti Astra (YDBA), PT Astra International, Tbk.

M. Riza Deliansyah, Head of Environmental & Social Responsibility Division, PT Astra International, Tbk.

Kartini Mulyadi, SH, Senior Partner of Kartini Mulyadi & Rekan.

Mas Achmad Daniri, Chairman National Committee on Governance, Chairman Mirror Committee on Social Responsibility, ISO 26000.

Chrysanti Hasibuan-Sedyono, Vice Chair of the Indonesia Business Links Board of Management.

Elmar Bouma, Director of the Indonesian Netherlands Association.

F. Antonius Alijoyo, Senior Advisor of PT APB Indonesia, board member of Indonesian Institute of Audit (IKAI), Board member of Indonesian Professionals in Risk Management Association (Indonesia PRIMA)

Subarto Zaini, Founder of Center for Corporate Leadership and the Indonesian Society of Commissioners.

Drg.Sadrach, former Secretary-General of the Indonesian Dentist Association (1975)

Meuthia Ghanie-Rochwan, Expert in Organizational Sociology, University of Indonesia and Activist for the Improvement of the Framework of CSR.

Mahyudin Lubis, Director, PT Indo Tambangraya Megah, Tbk.

Tri Harjono, Assistant Vice President Corporate Communication and Corporate Social Responsibility, PT Indo Tambangraya Megah, Tbk.

Most of all, without the continuous support from my family, my husband Urip and my two children, Tya and Tinggar, I would not have been able to gain all the experience through my years working for Unilever, Indonesia. Particularly to Tya; thank you for all the emotional support and constant encouragement throughout the writing of this book, especially at times when I did not feel like I was able to continue and finish the book. And last but not least, my love and thanks to Pandya and Chandrika, who were always around to help me whenever I had problems with the computer.

Benefits of Corporate Social Responsibility

Collective corporate social responsibility (CSR) activities amongst various corporations and its stakeholders could contribute to the microeconomic development of a developing country through *sustainable benefit to all.* At the same time, *optimum national impact, cooperation, and communication should be encouraged and socialized.*

Government

- Development and acceleration of microeconomic sustainable growth through using "good corporate governance/value change" and "best practices," resulting to a market environment conducive to both local and foreign investors (with the availability of good infrastructure, good education and health facilities, well-trained human resources and labor, and well-cared-for environment)
- Encouraging CSR activities, giving benefit to the community, with meeting certain development and sustainability criteria possibly being considered for tax incentives
- Joint CSR budgets possibly representing an additional source of public revenue (employment and wealth creation to reduce poverty)

Local community and society

- Changed habits, improved quality of life
- Capacity building, creates employment and wealth

Corporations

- Growth, profit, image, and competitive edge
- Community acceptance and goodwill
- Pride and spiritual values to employees and their families
- Genuine dialog with stakeholders

The world and environment

- Waste management
- Balanced ecosystem
- Green and clean environment

Preface

An interesting article was posted on journalism.org. This is a site managed by the Pew Research Center's Project for Excellence in Journalism, an independent organization working on projects dedicated to help people understand the information revolution.[1] The article discussed the George W. Bush and Al Gore presidential election in 2000 and how the press reacted to the final stages of campaign during the week of candidates' debates. The site stated:

> What also emerged, in debate coverage and elsewhere, was a clear predilection toward negativity. More than half of all stories contained twice as many negative assertions as positive. This finding stands out especially because during this election cycle, citizens have repeatedly told the media they are tired of the negative nature of campaigns. Further, studies of press coverage in general rarely find such great disparity between positive and negative coverage.

Although we are not using this specific example as a general liking toward a negative interpretation of events, in reality—that includes the business world—bad news does travel faster than good. With the advances of communication technology, the media, and the internet, customers share their views easily through their social networks for the whole world to know, and corporate behavior, whether good or bad, develops to become a global issue instantly.

During the writing of this book, a case in Indonesia grabbed the headlines of local newspapers, television, radio, and the internet. A young woman was charged for defamation and was sent to jail, awaiting trial after a criminal case was filed against her for sending an email to her friends. The email contained her complaint about the poor medical service she received at a hospital a few months earlier. This particular email went from her friends to various mailing lists. In the meantime, the hospital, seeing the damage the distribution of this email might cause to the company's image, filed a lawsuit against her. A decision that was meant to preserve public trust turned out to be disastrous news for the hospital when people perceived this as a threat against customers' rights. A heavy public outcry, built up by the internet social network and the media, later forced the authorities to release the woman from the penitentiary. She was put under house arrest. Finally, both parties agreed on a peaceful resolution, but the hospital had much to do to restore its reputation.

A small story like this becomes important in today's complex business world because reputation and goodwill is developing into the most valuable, and at the same time, the most vulnerable asset of a company. Goodwill is an intangible asset of a company; the loyalty of its customers or what is sometimes referred to as brand equity. The rising public expectation of businesses to act responsibly makes reputation worth far more than any tangible assets owned by the company because market capitalization vastly outweigh the real cost of hard assets such as factories and buildings.

Results of a global public opinion survey on the changing role of companies, the 1999 Millennium Poll conducted by Environics International, highlighted some valuable points on society's expectation of corporate behavior. The survey interviewed 25,000 citizens across six continents.[2]

- In forming impressions of companies, people around the world focus on corporate citizenship ahead of brand reputation or financial factors.
- Two out of three citizens want companies to go beyond their historical role of making a profit, paying taxes, employing people, and obeying all laws; they want companies to contribute to broader social goals as well.
- Half the population in countries surveyed are paying attention to the social behavior of companies.

- More than one in five consumers reported either rewarding or punishing companies in the past year according to their social performance. Almost as many would consider doing so in the future.
- Among those who expressed this view, about half defined the corporate role as "exceeding all laws, setting a higher ethical standard, and helping build a better society for all." The remainder said companies should operate "somewhere between" the traditional definition of the corporation's role and the more demanding one.[3]

These facts confirmed the shift of values that has driven companies to change their objectives from just "profit" to "profit, people, and planet"; a new attitude toward social obligations and the world we live in. Currently, businesses that are considered as being socially responsible are the ones that consciously target their business activities at these three-dimensional value creations of profit, people, and planet.

All through this book, these triple-bottom-line measures are merged into the landscape of corporate social responsibility (CSR), because profit forms the basis for the continuation of a company, and is a precondition to achieve the other two social dimensions. Thus, ensuring sustainable profits is mandatory, but only a holistic approach of all aspects of sustainability—a point where CSR plays a significant role—enables corporations to shape their competitive edge.

Many corporations realize that their second-most important asset is human capital. To develop their human capital, corporations go through a series of activities such as employment, training, education, capacity building, and other social support to individuals working for the company. These activities, which seem to be focused only on creating values and earning income for the corporations, are in fact social activities that indicate corporate responsibilities within their core business operation (internal CSR), with a high degree of control held by the corporations.

One of corporations' efforts to strive to get their human capital motivated is to set up clearly defined core values. Later in the book, we will discuss corporate core values from the perspective of good corporate governance. Businesses are expected to contribute to society's prosperity in the longer term by satisfying the needs of the community and to establish quality relationships with its various stakeholders based on good corporate governance and, at the same time, ensuring the corporate focus on environmental safety and protection.

Basically, this book concentrates on external CSR, which is the interaction of the corporation and the external environment outside the core business operation.

Over time, companies understand that sustainable growth of earnings and profit would not be possible without businesses continuously increasing consumers' demand or conditioning the market to accept products or services. In this book, a specific case is Unilever in Indonesia, whose CSR strategy of improving a community's quality of life through community development and education has helped it expand its market. This company has been able to maintain high consumer loyalty for decades and increase its competitive edge by continuously recognizing the changing needs of its now "educated" consumers, identifying gaps in the market and developing innovative products that consistently meet those changing needs.

More cases of Indonesian and multinational companies: Danamon, Astra, Pura, Indo Tambangraya Megah (ITM), Heinz, TNT, Intel, and Motorola, are discussed to show that application of CSR principles with the right understanding of community needs benefits both sides and truly provides these companies with a long-term competitive advantage.

We are going to learn from some of these companies that establishing an effective value chain, with both backward (raw and packaging suppliers) and forward linkages (sales distributors and retailers), by investing in long-term development and partnership with small and medium-sized enterprises (SMEs), ensures a steady and reliable supply of high-quality input for the business. It also provides opportunities to employ business ethics, apart from the more operational sides of technical assistance, training, or quality assurance; and although it is not directly visible, it builds respect for these companies resulting from employment numbers and community wealth created from these activities.

There are also cases where business supports contributions to the community and environment either directly, or by employees during their free time. This is an effective way to retain capable human resources, because such an action satisfies the spiritual value of employees and increases pride and loyalty, thereby making these companies preferred employers. Within the current global competitive market, the ability to keep good human capital is a clear competitive edge.

Companies selected as models presented in this book have recognized the need to preempt and avoid negative public perception by carefully nurturing and protecting their reputations and goodwill by doing the right and smart corporate social responsibility activities. Although the basic principles of CSR are similar all over the world, successful CSR implementation must be tailored to the needs and the developmental stage of a given country. It is my hope that this book provides some useful insight into mutually beneficial corporate social responsibility practices in the light of global competition.

NOTES

1. http://www.journalism.org/.
2. Environics International, The Prince of Wales, *Executive Briefing, The Millennium Poll on Corporate Social Responsibility*.
3. Jeffrey Hollender and Stephen Fenichell, *What Matters Most: Business, Social Responsibility and the End of the Era of Greed* (Random House Business Books, 2004).

Introduction

Corporate social responsibility or "CSR" is seen by some as merely the latest in a long list of academic management theories that can be ignored because it will soon be superseded by the next hottest fad of a creative and self-serving consulting industry. Many see CSR as an imposition placed on an organization by external parties, who believe that it is a responsibility that it should bear as part of its right to exist, in much the same way as a tax. They make reluctant contributions when they feel they have to. Still others see CSR as an altruistic obligation, morally imposed on successful companies as a sort of dividend to the community that has helped them achieve their success. They typically make one-off philanthropic gestures to mark specific successes.

These views miss the real point that CSR is a fundamental creed and methodology, which serves not only the wider environment but the best interests of the organization itself. It is a doctrine of "enlightened self interest," in which the organization, in the pursuit of its own success, helps the wider community create further opportunities that not only fuel the company's long-term success, but create ongoing benefits for that wider environment.

I believe that CSR is a basic business philosophy that should be a fundamental and integral part of the operation of every business enterprise. Not only is it the right thing to do, it results in substantial growth opportunities for those companies that fully embrace its principles. The ongoing success of a company depends on its recognition of its interdependence with its environment. In its most basic context, a business must have a market to sell to and the means of supplying to the needs of that market. It serves its own best interests to do all that it can to develop its opportunities to expand and enhance the wellbeing of the market as well as to improve its sources of supply.

At the earliest stage of my time as Senior Brand Manager/Marketing Manager of Dentifrice & Shampoo with Unilever Indonesia, I realized that I was already practicing the basic principles of CSR—although, at the time, it was not identified by its trendy acronym. In fact, the philosophy was an integral part of the company's strategic framework.

My objective in writing this book is to share my experience in pursuing "market development" through community building and education, improving lifestyle, and "gaining competitive edge" by creatively innovating products, packaging and services, tailor-made for a particular market segment, and establishing partnerships with small and medium-sized enterprises (SMEs). It is hoped that the widespread benefits of adopting the CSR philosophy will become apparent, and that more businesses and organizations will take on its principles.

In the 1970s, I was Senior Brand Manager of *Pepsodent* and *Sunsilk*; both are brand names for toothpaste and shampoo manufactured by Unilever (see figure I.1). In 1972, the total sales of *Pepsodent* was 959 tons, and those of *Sunsilk* 16,600 gallons. At the time, the total population of Indonesia was about 125 million, but users of toothpaste and shampoo were few, especially in rural areas. People were still using ground bricks to clean their teeth and plain toilet or laundry soap for

FIGURE I.1 Image of packages of *Pepsodent* toothpaste 120 grams (4.2 ounces), 25 grams (1 ounce).

washing their hair. Our efforts to develop the market through wider distribution and traditional advertising campaigns, using conventional media such as radio, TV, and newspapers, did little to increase the number of users and nothing to increase the frequency of usage. The expansion of the total market was only minimal.

It was recognized that efforts to increase sales was being severely limited by two major factors—the usage habits of potential consumers and the extremely limited purchasing power of most of the population. In response to the first part of the problem, it was realized that we could only be successful in increasing the sales of *Pepsodent* and *Sunsilk* if we could convince the people of the need to use toothpaste for brushing their teeth and shampoo for washing their hair.

The first step was to implement a plan of "community education" with the intent to change habits and create needs by emphasizing the positive benefits of using toothpaste instead of brick and shampoo instead of laundry soap. Now it is perhaps considered a clear-cut plan, but it was not an obvious idea then. So, in 1975, a joint initiative with the Indonesian Dentist Association was launched. The same collaboration with all dentist faculties in Indonesia was established, and educational demonstrations to emphasize the importance of oral hygiene were carried out at primary schools and in suburban and rural areas by young dentists, who gave demonstrations on how to brush teeth properly and offered free dental checkups and treatment.

After the success of the effort to educate the community on the importance of oral hygiene, a similar initiative was established in 1976 with hairdressers, who educated the community in the methods and advantages of washing hair with shampoos to get smooth, healthy, and shiny hair. This program emphasized the functional benefits of shampoos over laundry soap. These educational programs were supplemented by hairdressers' contests and the annual "*Gadis Sunsilk*" *(Sunsilk Girl)* contest, conducted in collaboration with the Indonesian Hairdressers Association and *Majalah Gadis*, then a newly published teenage magazine, which highlighted the emotional beauty benefits of shampoo usage.

The funding for these market building activities was categorized as "unconventional media budget", which was regarded as part of the advertising budget, and handled by Unilever's advertising agency in cooperation with professional organizations in their particular fields, that is, the Indonesian Dentist Association, the Hairdresser's Association, and so on.

Moving on to the second problem, it was acknowledged that within a developing market, limited purchasing power was a major impediment to market growth. This was especially so in the poorer class group and in rural areas, where earnings were traditionally received daily or weekly. This lack of purchasing power meant that the traditional developed-market method of stimulating sales by discounting prices of larger "economy" packs was not appropriate for the large majority of the Indonesian market. Potential consumers with limited purchasing power did not, at any particular point of time, have the money available to buy the larger size packs, and were therefore effectively excluded from the market.

The importance of the large rural and poorer segment of the market was well recognized and the solution was seen to be the introduction of good quality product in low-unit-price packs. Although the price per unit weight or volume was higher than it was for the bigger-sized packs, the smaller-sized units were affordable to those people with limited purchasing power. Unilever introduced small *Pepsodent* tubes of 25 grams (one ounce) and shampoo sachets of 12.6 milliliters (0.4 ounces) to encourage trials, to cater to first-time users, and to give an avenue for this particular type of consumer to sample the product. If you are wondering about the odd figure of 12.6 milliliters, it was actually the result of an in-depth research done to define the right single amount of shampoo to use for a person with long hair. During the 1970s and 1980s, most Indonesian girls loved long hair, and it was very important to determine the most cost-effective volume for the best result (see figures I.2 and I.3).

To motivate impulse buying particularly for first time users, the small shampoo sachets were displayed in attractive hangers. The introduction of these affordable unit-price packs facilitated the almost universal access to high-quality products and had substantially increased Unilever's markets and, at the same time, increased substantially the Indonesians' quality of life.

The development of this wider market would have been severely restricted had Unilever not developed its partnerships with SMEs to

FIGURE I.2 Popular hairstyle of the 1980s used in one of Unilever's posters, showcasing new *Sunsilk* black, a special product line formulated for black-haired Indonesian girls. The small simple plastic sachets were also introduced to the market.

supplement its supply capabilities. Presumably, the important thing in cultivating this partnership was an unreserved involvement through the provision of technical know-how and advice to those within the company's value chain. These would include material and packaging suppliers, third-party manufacturers, dedicated sales distributors, and suppliers of a wide range of services, and also the advertising agencies that were responsible for the brands' advertising.

FIGURE I.3 Low-unit-price pack of *Sunsilk* shampoo, containing 12.6 milliliters (0.4 ounce) of shampoo

In expanding the market, Unilever in Indonesia ran on a strategy of creative innovations on product, packaging and services, combined with creative and energetic marketing. These activities complemented the task of establishing new business partners by using management skills, technology, operational excellence, and business ethics. The outcome was twofold: a bigger market and more room for competitors. Nevertheless, the first company to develop the market for a particular product or service will benefit the most. In the case of *Pepsodent*, Unilever benefited from the development of a much larger market for toothpaste, while the community at large benefited by gaining a substantially improved quality of life through improved hygiene and health. The awareness and importance of oral care improved dramatically, especially in rural areas. Unilever had built considerable equity within the brand. Within five years, sales of *Pepsodent* had increased by 265 percent; from 2,303 tons in 1975 to 6,102 tons in 1980, and it had become the dominant market leader. It has been able to maintain its sustainable double digit volume growth, despite the many attempts of other international brands to enter the market. Today, *Pepsodent* is still the market leader, and has become so entrenched in the minds of consumers that it has become the generic name for toothpaste in Indonesia.

Similar to the case of *Pepsodent*, *Sunsilk* somewhat followed the same development pattern. People started to recognize the benefit of using shampoo; they realized that shampoo gave them smooth and shiny hair, which was not the case with toilet or laundry soap. Whereas *Pepsodent* toothpaste is a functional product, shampoo is more fashion oriented and Unilever needed to introduce other brands catering to the different fashion or trend to complement *Sunsilk* as the foundation brand. Within five years, sales of *Sunsilk* shampoo had increased by 690 percent from 64,000 gallons in 1976 to 443,000 gallons in 1981. Again, having developed the market through community education, Unilever's total shampoo range (*Sunsilk*, *Clear*, *Dove*, and *Lifebuoy*) continues to enjoy significant market leadership.

Its continuing success has enabled Unilever in Indonesia to further its philanthropic contributions to its environment through the activities of its care foundation known as "Unilever Indonesia (ULI) Peduli," which was established in 2000, and other community-minded projects. The embedded principles of CSR have been a major contributor to the ongoing success of Unilever, and have done much to increase the continuing wellbeing of all Indonesians.

From my observations, the iconic brands *Aqua* (bottled mineral water), and *Teh Botol Sosro* (ready-to-drink bottled tea) are in the same position as *Pepsodent*. They have worked hard to develop the market and established the value chain, so have been rewarded with significant market leadership. They will continue to maintain their strong positions despite competition from other brands that have copied their concept and entered the market.

Later inside the book, you will find more practical knowledge based on decades of experience in applying CSR principles, and you will also recognize that implementation of community-minded projects that had started more than 30 years ago, apparently, fit perfectly with the modern CSR theories. No statistics will be presented; only rich observations on actual markets. With the intention to highlight CSR implementation within a developing market, real case studies will show what really works for both the business as well as the community, described in great detail in the second part of the book.

For the sake of comparison, several exemplary models from developed countries are also included to show the obvious similarities of the principles. Anywhere in the world, sustainable growth and profit can only be attained through market development in an environment that is conducive for further expansion. Companies whose CSR programs are able to develop the market will always lead. Ensuring products or services have a competitive edge—by continually identifying consumers' changing needs, encouraging innovation, communicating and promoting through community building, improving lifestyle, caring for the environment, managing an effective value chain, preserving corporate reputation—is required to support market expansion. It is easy to see from the cases that this framework is applicable for companies operating in countries at various development stages.

PART 1

STRATEGIES

1

CORPORATE SOCIAL RESPONSIBILITY, GOOD CORPORATE GOVERNANCE, AND REPUTATION RISK

In general, people in business have yet to see the benefits of corporate social responsibility (CSR), and therefore have had no incentive to include the philosophy in their strategic framework or in their operational processes. Most still consider CSR activities as a sort of discretionary favor granted to the community by the business, and that such largesse is only appropriate after the company is well established, growing, and profitable. CSR of this type typically results in a one-off direct benefit to the community and very limited benefit to the company itself. Significantly, the benefits to both the community and the company are not sustainable.

On the other hand, driven by an ongoing revolution in communications technology, and underpinned by broader political, economical, and social changes, all businesses within a country are inevitably becoming part of the wider global market. Over the past 15 years, the world has also seen significant changes: the collapse of communism, the liberalization of China, Vietnam, and India, the emerging activities of the nongovernmental organization (NGO) sector, environmentalism, fundamentalism, consumerism, protectionism, World Social Forums, and so on. These changes have had a profound effect on not only the attitudes of governments and businesses, but also the people.

Advancements in information technology have also led to the availability of global television networks and the internet, which easily disseminate information instantaneously. Critics of business become better informed, helped by the global communication and the internet, while customers and consumers are better educated and becoming more aware of their rights and their potential power to influence corporate behavior.

Within this aggressive changing global marketplace, in addition to opportunities, businesses have also to face multiple challenges, as shown in figure 1.1. The impact of globalization has touched different business dimensions: from human resources aspects of capability and service competition to human resources development; from business operations to governments, NGOs, and consumers; and unquestionably CSR develops into one important area to deal with.

In acknowledging the complexity and diversity of the fast-changing world, and to ensure sustainable growth also to gain competitive edge,

FIGURE 1.1 Globalization offers both business opportunities and challenges

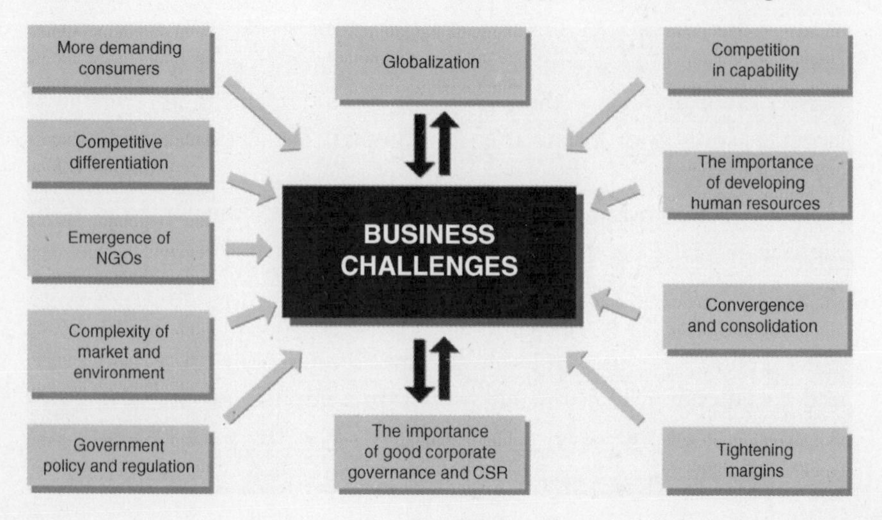

businesses globally have transformed their key business processes into strategic capabilities to facilitate the change from "product" into "services capabilities competition." Traditional functional marketing skills are necessary, but they no longer are sufficient. Broader-based leadership skills and capabilities, as a result of cross functional teamwork, are required.

At the same time, the mantra of business people has also evolved from "profit" alone into "profit, people, and planet." This new concept includes concern for several issues, those relating to people, to social issues, and to the environment:

- *People issues* range from workers' health and safety and employee morale, engagement, and development to corporate culture and good corporate governance.
- *Social issues* embrace community building, education and issues such as entrenched poverty.
- *Environmental issues* include concern for global warming, pollution, and disturbance of ecosystems.

All these factors are no longer considered as incidental, so CSR has come to the fore as a *core business issue*. One of the first organizations that got businesses to involve and respond to sustainability concerns is the World Business Council for Sustainable Development (WBCSD).

As a global association, dealing with the promotion of sustainable development, the WBCSD was looking for a generic definition of what CSR means to businesses because governments, the market, communities, and NGOs expect companies to do more, while companies demand more flexibility in CSR implementation. Fundamentally, the CSR–WBCSD journey supports the view that a *coherent CSR strategy based on sound ethics and core values offers clear business benefits*. In other words, acting in a socially responsible manner internally as well as externally is more than just an ethical duty for the company. It is something that actually has a bottom-line payoff, and at the same time could mitigate the corporation's reputation risk.

The WBCSD has advanced three concepts delineating the boundaries of CSR with the intention of suggesting to companies the most appropriate boundary of control and influence:

FIGURE 1.2 Spheres of influence adopted from World Business Council for Sustained Development (WBCSD) to accommodate the unique environment in Indonesia

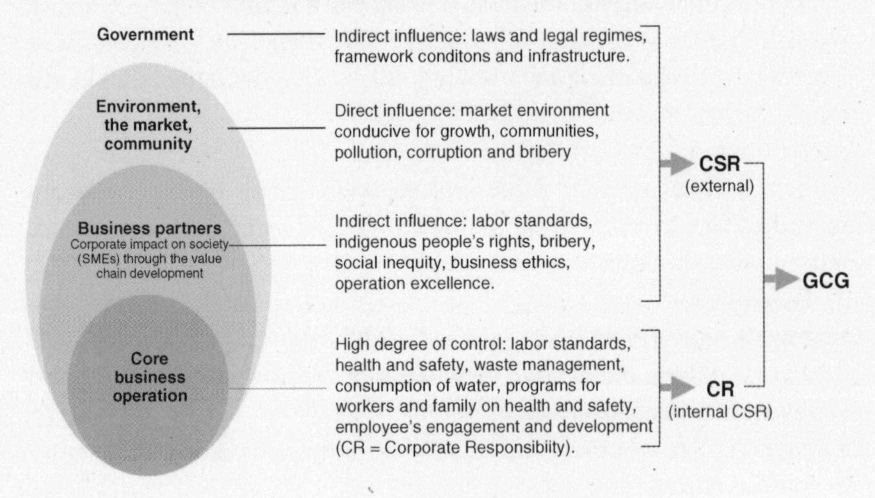

1. *Spheres of influence*

- To illustrate the decision and control inter-relationship between a company and its stakeholders, the WBCSD defines the CSR role of a company in spheres of influence. The diagram shown in figure 1.2 depicts the boundaries as nested circles of responsibility:
    - The inner core contains matters that are within the company's control, such as labor standards, health and safety, and waste management.
    - The outermost circle in which decisions and relationships contained are subject to the least amount of corporate scrutiny or influence.

2. *The value chain*

- To map issues and dilemmas along a value chain or a product lifecycle.

3. *Questions for the board*

- To identify corporate values and issues
- To analyze the impact on the value chain
- To communicate, do outreach, and influence

These questions can provide a valuable framework for businesses in defining "what's in and what's out" of the CSR box.

The WBCSD has found that the *CSR priority issues* today are *human rights, employees' rights, environmental protection, supplier relations,* and *community involvement.* Two related issues that cut across the others are the rights of stakeholders and the monitoring and assessing of CSR performance.

For our discussion, the WBCSD "Spheres of Influence" is adopted to map the author's experience in successfully implementing community-related programs in Indonesia that are now categorized under CSR.

By doing the mapping as in figure 1.2, we can define very clearly the boundaries of corporate responsibility (CR), which is internal CSR and external CSR. In summary, good corporate governance can be described as the incorporation of high standards of corporate behavior as a culture within the core business operation (CR) and in the interaction of the corporation and the external environment outside the core business operation (CSR). As mentioned earlier in the preface, this book only concentrates on external CSR, which involves the interdependent relationship of the business' core operation, its partners, the environment, and the market, including the wider community and the regulators (government).

Both good corporate governance and CSR are now becoming increasingly important parts of the business strategy. Good corporate culture and human resources capability are major determinants of a company's success or failure and CSR activities are important tools to support the company's strategy and image or reputation. It is only when a company is able to exert a high degree of control required to ensure good corporate governance within its core business operations that it can expect to successfully practice sustainable CSR for the benefit of both the company and society.

Successful internal CSR (CR) and corporate governance of an organization—with a well-understood vision or mission supported by a clear guiding principles, a demand-driven process organization, integrated human resources development, and an effective monitoring and control system, led by a strong exemplary leadership—will have "professional skilled human resources." Only professional skilled human resources can define and implement the right CSR in line with the business' vision and mission, and guide, lead, and negotiate, as well as train or change the values or mindset of business partners successfully.

Given the complexity and challenges, if a business is to survive and thrive, it must define the optimum balance among social, environmental, and economic factors for short- and long-term performance and profit. *This means that good corporate governance and CSR activities should be embedded into the company's culture and become an integral part of the short- and long-term strategy of the company.* To achieve this, and to ensure the sustainability of profit and growth, businesses should adopt values, principles, strategies, and practices that are both compatible and consistently applied.

Business need not be concerned that its motivations and intentions in adopting CSR within an overall objective of growth and increasing profits might be misinterpreted by a cynical public. The WBCSD noted that business and the society at large are interdependent and through mutual understanding and responsible behavior (good corporate governance), the business role in building a better future should be recognized by society.

Further, as Adrian Cadbury said in 2002, because unprofitable business is a drain on society, there is no conflict between social responsibility and the efficient and profitable use of scarce resources by business. The essence of the contract between society and business is that companies should not pursue their immediate profit objectives at the expense of the longer-term interest of the community.

Focus Point

A company can expect to practice sustainable CSR successfully only when it is able to exert the high degree of control required to ensure good corporate governance within its core business operations.

It is extremely important to ensure the ongoing support and understanding of the community because the business world has changed. Value creation to the business, as well as to the wider society, should now become the ultimate objective of any business, and adopting this concept would entail a fundamental internal change; that is, to integrate

CSR into the business processes. This has to be considered a strategic approach because people may have to recognize that CSR provides an effective mechanism to create, preserve, and improve the company's reputation and image. There are quite a few examples of companies struggling to withstand a damaged reputation, not because they have breached the law, but mainly due to the lack of trust and acceptance from the community, or the lack of company's understanding of the local community's needs.

With the high-speed development in communication technology and the media, bad corporate behavior can travel fast across the globe, delivering a hard blow to the organization, harming the company's reputation. However, companies with an approved contribution to the people and the environment—thereby gaining the trust and confidence of the community—will have a better chance to protect their reputation. *The company's impact on the community is an excellent catalyst for social accountability, and at the same time promotes positive image and reputation.* Unlike financial risk, which typically involves quantifiable market, credit, and operational risks, it is impossible to estimate reputation risk up front. But the cost of losing this most precious asset can be very dear.

CSR and GCG can be seen as intangible elements that contribute to corporate success, with equal importance but different perspectives from that of the financial measures. Plotting the chart in figure 1.3 to the spheres of influence gives us a clearer picture of the role of CSR and good corporate governance in safeguarding, or better yet, in improving the company's reputation.

Reputation reflects the perception, good or bad, of different groups of people who interact willingly with, or are affected by, the organizations that are in their sphere of influence. We usually call these groups stakeholders. They build their perception based on their evaluation of the businesses' culture, behavior, and performance, while working within or in cooperation with the businesses, or through available information, which disseminates very quickly in this new world of advanced information technology.

In a survey of 269 executives conducted by the EIU,[1] reputation risk emerged as the most significant threat to business out of choices of 13 categories of risk. The respondents also felt that risks to their

FIGURE 1.3 The links among nonfinancial risk measures management

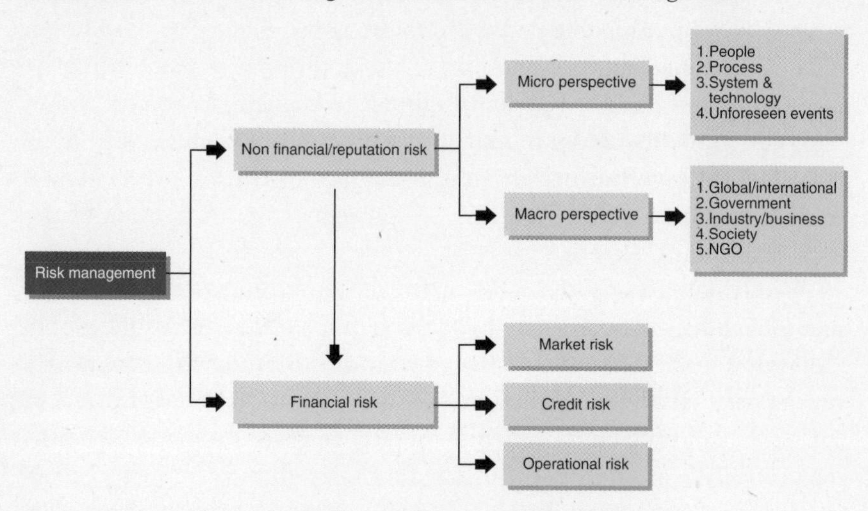

companies' reputation had increased significantly over the past five years. (See also Company's Challenges in figure 1.1)

Quoted from various sources, mainly from www.erm-academy.org by F. Antonius Alijoyo Board member of Indonesian Professional in Risk Managemennt Association (Indonesia PRIMA) as workshop materials; the following are several risks that are categorized as reputation risk:

- Reputation risk is the current and prospective impact on earnings and capital arising from adverse perception of the image of the company by stakeholders, regulators, and the public. This will affect the organization's ability to establish new relationships of services or continue servicing existing relationships. This risk may expose the organization to litigation, financial loss or a decline in its customers' base. Reputation risk exposure is present throughout the organization and includes the responsibility to exercise a lot of caution in dealing with its customers (good corporate governance) and the community (good corporate governance and CSR).
- Reputation risk is the impact of third-party (government, NGOs, local community) pressures and influence on the environment in which a company is operating. These are the externally imposed

limits on an organization's ability to operate within a particular jurisdiction or regulatory environment.

- Reputation risk is the risk that a latent reputation will become an actual reputational problem.
- Reputational problem is the risk that negative publicity regarding an organization's business practices will lead to a loss of revenue or litigation.

People, process, systems and technology, and operation are basically components of the core business that are within the close control of the company, and this is an area where good corporate governance plays a major part in building a good reputation. From the macro perspective, a business has little or no control over the government, industry, environment, society, and now, with globalization, the international community, *but well-defined and strategic good governance CSR activities will ensure the optimum balance among social, environmental, and economic factors for short- and long-term profit sustainability.* Well-managed corporate governance and CSR equal good risk management.

NOTE

1. Economist Intelligence Unit, *Reputation, Risk of risks* (2005).

2

A NEW PHENOMENON

Globalization has made the awareness of business responsibility to the society grow and corporate social responsibility (CSR) is *widely recognized as a worthy commitment to ensure sustainable benefit for both corporations and communities.* CSR is becoming an important base of a business to build the trust and confidence of stakeholders, and this could be the key source of competitive edge.

However, implementation of CSR activities to give optimum values for business and community has yet to become well understood. Projects such as a two- or three-day community environmental fair displaying concepts of energy savings, waste management, and other ideas concerning care for the environment, are often perceived as part of CSR initiatives. In fact, one-off activities like this without capacity building, training, empowerment, employment, and wealth creation will not result in a win–win situation, and consequent sustainability. Later, in part II of the book, you will find sustainable CSR programs in practice, for example, Unilever's Environmental Care Program (case V in case study 1) or Heinz's sustainable agriculture program (case I in case study 6).

If a given company stresses its CSR implementation on issues irrelevant to the local environment, even the best activities may fail miserably to create the intended benefit for either the community or the company. *The application of CSR principles would therefore directly be influenced by the business' understanding of its business strategy, in proportion to the need of the community.*

Corporate involvement in socially-related engagement normally considers larger economic, social, and cultural views of the country in which the business is operating. The situation in developing countries can be quite different than those of the developed countries. Although basically the principles of CSR are the same globally, each location calls for different emphasis in the implementation of CSR. Unlike in developed countries, where most states assume primary responsibility for social welfare of the community (after the World War II), it is not the case within developing countries.

International companies planning to optimize the opportunities of the wider global market by establishing its international supply chain or entering new emerging markets have to implement CSR-related initiatives following the rules of CSR activities within the developing market. This will ensure successful entrance, sustainable growth, and profit.

The US, European countries, and Australia are typically ahead in many social and community aspects of countries in Africa, South America, and Asia—with the exception of Japan. So a lot of CSR initiatives in these developed countries concentrate on missions that serve wider purposes, such as environmental sustainability. Having CSR as a "Western" product, it is not surprising that there is the perception that CSR as a business obligation deals only with environmental issues. However, people should realize that any problem concerning the environment is basically a result of the community's behavior.

Take the easiest and the most obvious example. Large cities in the developing world, with their ever increasing urban population, are welcoming modernity without the necessary infrastructure and community education required. We can easily see modern high-rise buildings, luxurious apartments, condominiums, and offices, also newly built businesses or manufacturing districts and big highways surrounded by filthy, overcrowded slums. Dirty rivers, where people pile up their household garbage, are ironically the primary water support for those living along the riverbanks, and also for the city.

Scarcity of clean water supply due to poor groundwater management, polluted soil, air, and water, are only a few of the apparent problems faced by these cities, not to mention the responsibility to meet the economic, sanitation, and other community infrastructure and services needs. Not only are people living in slums, but the city dwellers are also walking at a slower pace against the speedy development. They are being left behind, facing massive social, economic, health, and environmental problems.

For cities with complex issues, CSR programs must be designed very carefully for optimum benefit and acceptance. For example, the understanding of the importance of caring for living areas and of preserving the environment can only be brought about by changing the community's mindset and values through community development, followed by constant training in the changed value so as to become a habit.

In one of the case studies presented (see case V: Unilever's Environmental Care Program in case study 1: PT Unilever Indonesia, Tbk), you can find a CSR program, carried out by Unilever in Indonesia, designed to reinforce habit change in relation to environmental protection, starting from individual persons within the smallest household community. People living in dirty crammed alleyways behind handsome buildings and alongside polluted rivers, loaded with litter and human waste, started to learn the concept of a green and clean environment within and outside the house through waste management.

In collaboration with the local authorities, Unilever recruited local cadres and volunteers, and trained them in simple household waste management so that they could go on to train the community. The training provided them with basic knowledge of dividing waste into different categories, processing the biodegradable waste into compost and recycling plastic waste into products with economic value.

Because change of habit is a process, the program also offered a continuous monitoring system and support. This support was critical to avoid failure of the project and to facilitate the program to be expanded to other cities and areas.

In summary, for most of developing countries, where typically the education level of most of the population is relatively low, any CSR activity—not limited to those concerning the environment—have to start with community building, that is, to change the mindset and values of the said community. Community investment in developing countries has to accommodate the needs for basic and vocational education, infrastructure development, and wealth creation, while in the developed and industrialized countries, programs on more advanced education concerning technology and innovation are appropriate, aside from environmental sustainability-related programs, philanthropy, employee volunteerism, and other initiatives toward local communities.

So for the less economically developed nations, it is important that CSR activities put some focus on the creation of sustainable wealth for the community, particularly for the lower-income society. This can be done through community training and capacity building, technology leverage, health campaign, including employing operational excellence, good corporate governance, and quality assurance. *The overall socioeconomic betterment will consequently generate sustainable growth and profit for the company.* It is proven that the right application of CSR principles and formation of partnership create the most effective social impact.

The World Business Council for Sustainable Development (WBCSD) considers social investment to be increasingly seen as a necessary part of doing business, especially in developing economies that lack basic infrastructure and the capacity to build social capital. The commercial justification of such investments lies in their contribution to a healthy and stable business climate. This statement is exactly in line with the author's experience in Indonesia, as later described in the case studies.

The following results of a CEO survey conducted IBM (2004, 2005, 2008) also confirms the presumption. According to the survey, only three change drivers continue to rise in importance: socioeconomic factors, environmental issues, and people skills. All three of them are linked to CSR.[1] CEOs or companies who focus on CSR issues invest more in developing new "socially responsible" and "green" products and services. While customers have always cared about social issues, their concerns are now more frequently turning into action and influencing purchasing decisions (see figure 2.1).

The report clearly noted that CSR is shaping the future:

> With talent in short supply, employers' Corporate Social Responsibility reputations are an important tool to attract and retain employees. At the same time, companies are also recognizing that they are being held mutually accountable, along with the public sector, for the socioeconomic wellbeing of the regions in which they operate.

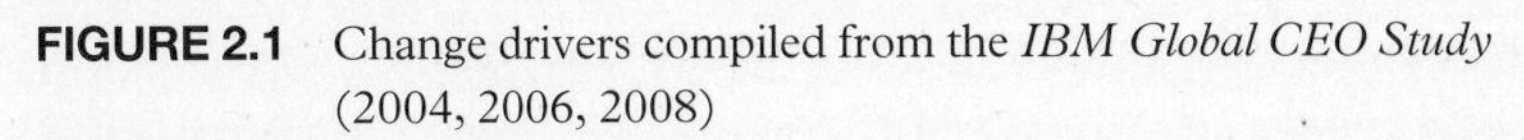

FIGURE 2.1 Change drivers compiled from the *IBM Global CEO Study* (2004, 2006, 2008)

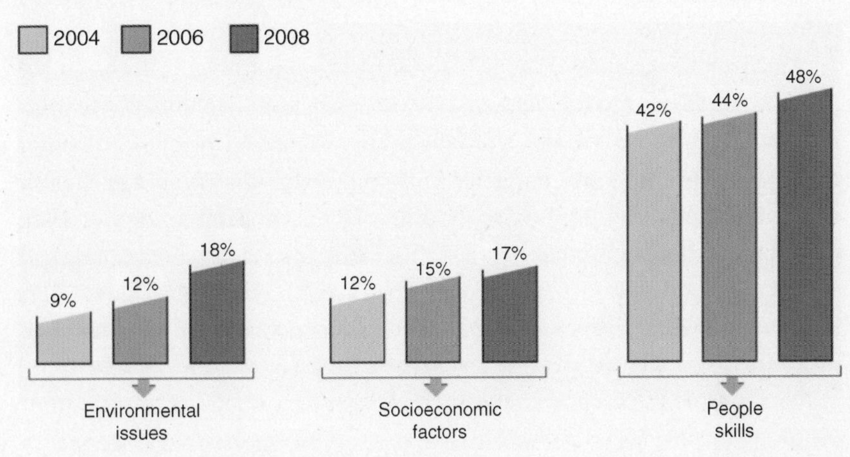

Though CEO concern about environmental issues has doubled over the last four years, but this concern are not evenly distributed. *Asia Pacific is taking the lead with a dramatic increase in interest, nearly tripling since 2004*, followed by European and America's CEOs.

This phenomenon is easily understandable. Businesses and CEOs in Asia–Pacific are used to developing the market through community education and improving lifestyle and capability development (social impact on the value chain) to ensure growth within this developing market. Therefore, businesses and CEOs in Asia–Pacific only have to expand their community activities to include environmental issues.

It is also unsurprising that CEOs are generally positive toward the rising of CSR expectations and they are seeing opportunities:

> Regulatory compliance is considered less of a problem compared to facing media and political pressure from socially active environmental NGOs. Exposures of this kind could certainly have a negative impact on the *business reputation* and *image. CEOs/Corporations are mostly aware of their obligation to "do no harm" and are conscious of the emerging generation of socially minded customers, workers, partners, investors, regulators, and*

non-governmental organizations (NGOs) who are watching and monitoring every step companies make.

On the other hand, they also see opportunity to build competitive edge through having a smart and clever CSR plan, specifically to gain market share and maintain sustainable growth and profits. Corporate identity and CSR are becoming important differentiating factors of brands and companies, especially when entering new markets; described in the report as:

> Consumers will increasingly make choices based on the sources of the products they buy including the ingredients and processes used in making these products.
>
> —*IBM Global CEO Study (2004, 2006, 2008), page 63*

More than 70 percent of the world's population lives outside the industrialized world, and accordingly, makes up the majority of consumers, and at the same time, the major resource of outsourced products and services. It is therefore worth noting the specific needs of different developing countries as a guide to an effective CSR program.

One important point to note is that in many developing countries, corruption has become a major problem. In a paper developed by the Partnership for Governance Reform (2000), an organization committed to promoting good governance in Indonesia, the main issue with corruption deals with the mindset that tolerates corruption, the inadequate governmental system, and the lack of control and enforcement. The key strategy for fighting corruption is taking the human resource as the foundation. People need education on attitudes that reject corruption, and also need to be equipped with proper skills, competencies, and capabilities that help them build anticorruption behavior. For a country to effectively eliminate this rooted habit, this anticorruption strategy takes a building-block approach, which considers the whole governmental system to support the human resources development (see figure 2.2).

The paper mentions that control and enforcement are the most immediate components required for other components to work to avoid rampant abuses of the law, political, and economic system.

FIGURE 2.2 Fundamental building blocks for an anti-corruption strategy

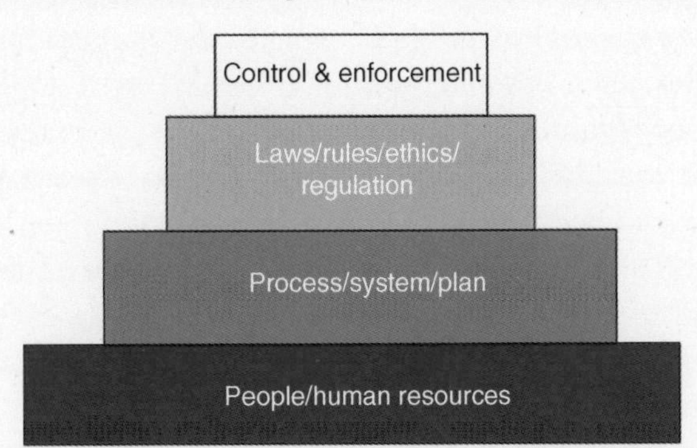

Effective implementation of CSR in developing countries that have weak law enforcement, occasional financial and political instability, poverty, and a low educational level, requires also the machinery of an effective democratic government and civil society.[2] CSR is not a tool to solve issues of poverty, lack of work opportunities, exploitation, and environmental devastation, but rather an instrument for corporations to help the government by supporting and empowering communities and local structures.

Basically, building a nation is not different from the establishment of a corporation. From the beginning, the government has to target its activities consciously at three-dimensional value creation:

- successful and sustainable growth of the micro and macro economy of the country (profit)
- ensuring society's prosperity in the longer term by satisfying people's needs in terms of provision of infrastructure, good public services, and education, as well as ensuring socially responsible investments in creating jobs and wealth (people)
- ensuring conducive environment through continuously caring for the community and environmental safety (planet)

To ensure successful results, strong control and law enforcement over these three-dimensional value creation processes are needed to change the vision to become a habit, leading to sustainability. Nevertheless, the progress of the development of each country depends very much on the history and its culture, and it is worthwhile to elaborate slightly on the different types and characteristics to help build a better understanding of the various countries' needs.

Developing countries can be divided into three main categories. However, different definitions may classify some of these countries into emerging economies or newly industrialized countries. The World Bank uses the gross national income (GNI) per capita, or previously the term was referred to as gross national product (GNP), as the main criterion to classify the world's economies and divide the world's countries into four income groups.[3] All low- and middle-income economies are categorized as developing countries with a note that this classification may not necessarily reflect the development status.

Developing countries are distributed mainly in Asia with the exception of Japan; Latin America, and small part of Europe; Middle East and Africa; while most of the least-developed countries are in Africa. More than 70 percent of the world's population lives in the developing and the less-developed countries.

The first type of developing nation is a country with a relatively small territorial area, a small population, and limited or lacking natural resources. Because of the nature of the territory, people cannot depend on agriculture or other land resources as their primary source of livelihood, but instead concentrate heavily on high skilled manufacturing and services. The first group of countries of this type is Singapore, Hong Kong, Taiwan, and South Korea, which has moved to newly industrialized country category during the past few decades.[4]

The historical background of these countries, specifically Singapore and Hong Kong as former British colonies, play a significant role in shaping the countries to rapidly adopt modern government and economic systems. In the early nineteenth century, the British Empire had made Singapore its commercial and military base for Southeast Asia, and Hong Kong its entry port for Asia.

To attract foreign direct investment (FDI), these countries put a major focus on the development of human resources capital through education, training, capacity building, and development of professional skills. Investments are also carefully selected, with priority given to the socially responsible investments (SRIs), which include investments in infrastructure: roads, airports, harbors, and public housing. This has a multiplying effect to attract more FDI and local investors, which in turn creates employment and wealth in the local community, and, therefore, provides sustainable growth potential.

In encouraging discipline, a fundamental component for country's development, government exerts control and law enforcement to ensure that values are becoming a habit. In Singapore, in the early days of its independence, the government launched the first nationwide public program in 1968 to keep the country clean. An article found in Singapore National Library mentioned that the government believed that improved environmental conditions enhance the quality of life of the citizens, cultivate national pride, and attract foreign investors. Under an antilittering law, violators were imposed hefty fines, which proved to be an effective tool to create new habits for Singaporeans. This first campaign marked a journey to empower the community to take greater responsibility in preserving the environment.

For countries with growing productive capital, an established financial system, and human resources with a better educational level such as these newly industrialized countries, the design of CSR programs can follow those that are applied in the developed countries, which of course requires adjustments to local needs.

The second type is the developing countries that are of significant interest to investors because of their strategic geographical position and the richness of resources. These are countries such as Indonesia or Nigeria, and in a different perspective, India. Developing countries classified here have large territorial areas, historically are agriculture based, with high population density, and lack self- and mass-discipline as a culture, which leads to ineffective development in many sectors.

In the late nineteenth century, several of these countries, especially those under Western colonialism, started the exploitation of their rich natural resources; and the sweet price of mining products swiftly shifted their economies from the agriculture sector as the main source of income. Large oil reserves were found, as well as other fossil fuels and minerals, such as tin, gold, copper, bauxite, and nickel. Indonesia and Nigeria are two examples of countries of this second type becoming important mining countries. They have neglected agriculture-based products, especially Nigeria, and to a lesser extent, Indonesia has followed the same pattern.

India in the context of development to some extent is unique, because it has realized that one of the major contributors to the country's economic development is the good education of its human resources. British colonialism has also brought the English language to become an important and commonly spoken language in India, which has given the country certain advantages in global business.

Tata Group, India's largest multinational conglomerate, has realized very early the importance of education as a driver of change and economic development. The company was established back in the 1800s, and even at that stage, Tata was granting scholarships for Indians to study abroad, and built the country's first science centre and atomic research centre.[5] These along with other community-based activities—although at the time was not labeled as CSR—has always been part of Tata's business operating principles. This is only one example, but it results in huge economic potential expansion from improved education (people) and the exertion of CSR principles within the business strategy, because currently Tata has operations in every major international market.

It is worth mentioning Thailand in this category. Unlike all its neighbors in Southeast Asia, this country was never colonized. Thailand's distinct agricultural history has long enabled the country to become the world's largest exporter of rice, and today, many other agricultural products. His Majesty the King of Thailand introduced a philosophy called the "Sufficiency Economy," which was adopted by the government to ensure food sufficiency at the local and national level.[6] The King, who is deeply loved by his people, has been the architect and the inspiration behind the development of Thailand's sustainable agriculture, including agricultural research and the management of the social, economic, and environmental impacts emerging from this important activity.

We are now going to look specifically at Indonesia, a country with a rich and also complex historical, cultural, and social heritage. Indonesia is a formerly agriculture-based country. The country, located near the equator, has only two seasons in the year, with no extreme seasonal temperature changes. Plus it has very rich soil, somewhat affecting the traditional attitude to farming. People are used to very hard work during the planting and harvesting time, and relax between the two busy periods. Traditional farmers, who constitute most of the population, have no entrepreneurial drive to do their own marketing and selling.

The government track of development moved into full industrialization of all natural resources. Foreign investments in the manufacturing sector increased; new plants were built to boost production of palm oil; and the pulp and paper industry took an important role in the country's economy to complement the earlier-developed oil and mining sector. People started to overlook the traditional agriculture sector, and farmers changed job to become cheap blue-collar laborers, entering the industrial world without proper training and therefore with low efficiency.

Indonesian developments are very much influenced by three main factors:[7]

- the historic and cultural legacy, which inhibits significant changes in corporate social and environmental performance
- the perception of good governance or code of conduct
- the environmental impact of large businesses with particular reference to the mining and palm oil industry

Indonesia was under Dutch colonization for almost 350 years, and the Dutch East India Company ruled most of the nation's economy. As a result of this early history, capitalism is closely linked with colonialism and imperialism in the Indonesian consciousness.[8] The legacy left by the Dutch colonial regime is a complex mixture of patronage, monopolies, and the concentration of wealth and power to well-connected group of businesses, linked to the Indonesian political elites; all of which encouraged the emergence of Indonesian conglomerates.

Between the end of the colonial era up to 1966, Indonesia's first president, Soekarno, was very adverse to foreign investment,[9] and by the end of 1966, any remaining Western capital had been repatriated or expropriated.[10] When President Suharto succeeded President Soekarno, foreign investors were encouraged to invest in Indonesia, and Unilever was one of the first to come back. By early 1970, Unilever Indonesia was fully operating. A lot of discussion about this company can be found inside this book because *Unilever Indonesia is the only multinational company in Indonesia that applied CSR principles at a very early stage of its business re-establishment* (from 1975 onward).

With years of political instability hampering the country and its economic development, Indonesia was 25 years behind its closest neighbors in its economic development after gaining independence. To accelerate development, investments in exploration and exploitation of the country's natural resources were offered to foreign investors. While the availability of rich natural resources was indeed very good for economic growth, the government was not equipped at that time with rules and regulations concerning mining and harvesting forestry products. The law on AMDAL (the Environmental Impact Analysis) for corporations operating in Indonesia was enacted in 1982, but only implemented after the government's regulation on the definition and implementation of AMDAL was issued in 1986.

At the same time, the government did not demand investors to build processing plants to create value-added products that would give the country longer-term profit.

Indonesia became an exporter of raw materials, which have much lower value than finished products. Although exploitation of natural resources did give direct benefit and profit to the nation, the people did not have the opportunity to fully share the wealth through the creation of employment by having local processing plants.

The financial crisis years of 1997–98 taught companies in Indonesia valuable lessons. Big companies and powerful conglomerates were severely hit by the collapse of the national currency, and they suffered huge losses. Those difficult years had made companies realize that long-term sustainability is not only based on high profit, but also on the availability of a local market environment conducive for growth, diversified consumer bases within and outside the country in which the company is operating. Especially for Indonesia, with its 230 million population, companies now also realize that ensuring a market environment that is conducive for business is not solely the government's responsibility, but also the responsibility of all big corporations.[11]

After this development, many big companies in Indonesia have adopted the principles of CSR in developing simultaneously the 3Ps: profit, people, and planet. First, these companies began operating as islands of integrity, where they started to adopt good governance in their internal operations and, at the same time, began to implement strategic and well-chosen strategic CSR activities, giving an optimum balance among social, environmental, and economic factors for long-term sustainable growth and profit.[12] This approach proved to be effective to mitigate the effect of the recent global crisis (2009) as Indonesia's GDP is still growing at 4.5 percent, compared to the negative growth of the neighboring countries, for example, Malaysia, Thailand, and even Singapore.

The third and last type of developing nation is countries with large territorial size, plenty of natural or agricultural resources and typically large population, and the most important thing here is that they inherited a culture of mass discipline. In this category are countries such as China.

We can learn from China's achievements in improving the country's development indexes. The communist history in China provided a good base for good governance, law enforcement, and elimination of corruption. With the Chinese economy opening up, these are the qualities required for a successful transition from the former state monopolistic system to a freer socialist market economic system.

According to China's Ministry of Culture website,[13] the nation's economic system reform started in the rural areas back in the 1978, and in six years, the restructuring shifted from the rural areas to the cities. The Chinese government has established a set of policies to support the reform. The main aspects of the economic structural reform based on the official document as cited on the site are as follows:

> The development of diversified economic elements will be encouraged while keeping the public sector of the economy in the dominant position. To meet the requirements of the market economy, the operations of state-owned enterprises should be changed so that they fit in with the modern enterprise system. A unified and open market system should be established in the country so as to link the rural and urban markets, and the domestic and international markets, and to promote the optimization of the allocation of resources. The function of managing the economy by the government should be changed so as to establish a complete macro-control system mainly by indirect means. A distribution system in which distribution according to work is dominant while giving priority to efficiency with due

consideration to fairness should be established. This system will encourage some people and some places to become rich first, and then they may help other people and places to become rich too. A social security system, suited to China's situation, for both rural and urban residents shall be worked out so as to promote overall economic development and ensure social stability.

In less than three decades, we can see that China has accomplished great results in many sectors, and now is in the fast lane to economic and industrial development, also to improvement of public health and education, and awareness of environmental sustainability.

NOTES

1. "The Enterprise of the Future", IBM Global CEO Study (2004, 2006, 2008). The study is based on conversations with more than 1,000 CEOs and public sector leaders worldwide.
2. Melody Kemp, *Corporate Social Responsibility in Indonesia, Quixotic Dream or Confident Expectation?* (United Nations Research Institute for Social Development, 2001).
3. http://web.worldbank.org/WBSITE/EXTERNAL/DATASTATISTICS/0,contentMDK:20420458~menuPK:64133156~pagePK:64133150~piPK:64133175~theSitePK:239419,00.html.
4. http://www.infoplease.com/cig/economics/world-economies.html.
5. http://www.tata.com/company/Media/inside.aspx?artid=ZYTmDvfrGd4.
6. http://thailand.prd.go.th/view_inside.php?id=4574.
7. Kemp, op. cit.
8. Wibisono (1991) as cited in Kemp, op. cit.
9. Hill (1991) as cited in Kemp, op. cit.
10. Hill (1996) as cited in Kemp, op. cit
11. This situation has been accelerated further by the development of the competitive global market development and the emerging of new industrial countries such as China and Vietnam, which are becoming very attractive sourcing countries or markets for new investments.
12. See the Indonesian case studies in part II of the book: PT Unilever Indonesia, Tbk; PT Astra International, Tbk; PURA Group; PT Bank Danamon, Tbk; and PT Indo Tambangraya Megah, Tbk.
13. http://www1.chinaculture.org/index.html, http://www.chinaculture.org/gb/en_aboutchina/node_71.htm.

3

THE FUNDAMENTALS AND EVOLUTION OF CSR

THE FUNDAMENTALS

Before the twenty-first century, there was a common perception within and outside the business world that the company's sole social responsibility was to make as much profit as possible, while community building and public services are the sole responsibility of the government. Therefore, maximization of profit meant the maximization of the taxes paid by the company to the government. This, in turn, could be spent on welfare, improving the society's wellbeing. Even now, almost a decade through the new millennium, we still can find similar opinions expressed in corporate discussions, although a different insight of business sustainability has started to be widely accepted.

With this traditional understanding, involving the company in corporate social responsibility (CSR) would be seen as detrimental to both the company and society in general. From the company's point of view, funds spent on CSR reduced profits and provided, at best, only minimal short-term benefits. From the government and society's viewpoint, the cost of CSR reduced taxable profits and therefore the taxes that could be used to increase the wellbeing of society. None of the parties considered CSR in any way related to their respective business or social activities.

At its most fundamental level, a business needs a market and a product. If it is to survive and subsequently thrive, it must find its market, then do all that it can to expand it by providing socially innovative products and services that satisfy the market's needs at affordable prices. Sustainable growth and profits require ongoing efforts to increase available markets for products with a perceptible competitive edge because a product's competitive edge changes with time, market environment, and perception of consumers. *The reward for success is market leadership in an expanding market.*

The Market

Sustainable growth can only be achieved by increasing consumer demand or conditioning the market to accept the company's product or services, by either increasing the number of users or increasing the existing consumption by those consumers. The product can be given room to grow if market expansion is stimulated by appropriate community development or education and other efforts to improve lifestyle. While efforts to expand the market will benefit all brands, specific market history shows that market leadership almost invariably remains with the brand responsible for the initial community education.

The Product or Service

To thrive, a company must develop products and services that have a discernable competitive edge. This can only be achieved if there is an ongoing commitment to recognize the changing needs of the consumers, to identify gaps in the market, and to develop socially innovative products that consistently meet those changing needs. Opportunities must be reinforced by continuing to communicate and promote the product and its benefits not only through the conventional media (TV, radio, newspaper or magazine, billboard, and so on), but also using unconventional media to promote community development and education, and to change habits and improve lifestyles.

Ensuring the most cost-efficient means of product supply is an important factor in establishing and maintaining competitive edge. This can be done through developing business partnerships with independent third-party small and medium-sized enterprises (SMEs).

Ideally, SME partnerships are established *at the earliest stage of a business' development*. This is the stage for a business to plan and define the location to manufacture its products to ensure the most effective logistics cost for distribution, and also which products to be produced in-house or outsourced to third-party SMEs. The development of SME partnerships need not be restricted only to production capability. It may be appropriate to establish SME partnerships to cover the provision of different services, for example, suppliers, logistics, sales and distribution, or promotion activities.

To ensure sustainable growth and a competitive edge, the need to develop a sales and distribution network through partnering with SMEs located in all important trading areas is very important.

The business helps the development of the SMEs by providing advice, guidance, training, and expertise in a number of different areas: technology transfer, employing best practices in the development of suppliers of raw and packing materials, quality assurance of processes and products, distribution and sales networks and control, expertise covering for instance simple accounting system, and quality assurance and technical return reporting to monitor production efficiency.

The relationships between the business partners should be formalized by the signing of codes of business principles (good corporate governance). All parties share substantial benefit from the establishment of appropriate SME partnerships. The reward to the company for the investment in the development of SMEs is the means of ensuring capital efficiency and frugal investment, along with cost effective and reliable supply of quality products, which ensures competitive edge and, consequently, growth within an expanded market.

The SMEs themselves not only benefit from the help to establish substantial cost-effective businesses but, perhaps more significantly, the potential to share long-term profitable growth as the indirect result of the developing business' ongoing efforts to promote sales and further expand the market. Society at large profit from a wide range of benefits, including the improved living standards resulting from the availability of the reliable supply of affordable good-quality products, the provision of more efficient services, employment opportunities occasioned by the expanded businesses and the promotion of a range of ethical, social, and environmental standards through the acceptance of codes of business principles.

Community building, education, transfer of technology, and the development of SMEs, which are all categorized under the area of CSR, have resulted in a substantial socially responsible contribution to the nation as a whole.

Unilever India (Hindustan Lever) and Unilever Indonesia have incorporated these policies into their strategic plans as early as the 1950s and 1960s for Hindustan Lever and Unilever Indonesia since the 1970s and 1980s and have continued with them into the present.[1]

As for Unilever Indonesia, as a result of the author's direct involvement in the adoption and implementation of socially responsible principles within the company, a more accurate observation can be shared here.

During these few decades, its turnover has increased from IDR16,081 million ($38.7 million at $1 = IDR415) in 1973 to IDR12,500,000 million ($1,390 million at $1 = IDR9,000) in 2007. Unilever started with developing the community, which gave it a much expanded market and, for most of this time, the company has maintained the same business partners, although some are now run by the second or even the third generation of the original partners.

Besides developing the market, in ensuring growth of the business Unilever Indonesia had also introduced locally their global brands (1970s and 1980s) and innovative new products and packaging. Many of these products were introduced in the Indonesian market for the first time, for example, branded deodorant (*Rexona*), the first detergent powder *Rinso*, laminate tubes for *Pepsodent* toothpaste to replace aluminum tubes, which was believed to be a contributing factor to Alzheimer's disease, laminated sachets for lower-priced packs to replace the PVC sachet, which was difficult to open, and so on.

Unilever Indonesia had also initiated the development of local suppliers of raw and packing materials to substitute imported products, and to avoid paying unnecessary additional costs of import duty and shipping. This was a very important decision, because by doing so, Unilever Indonesia had been able to sell high-quality global products at a price affordable to the local consumers.

Indonesia is a vast archipelagic nation, as can be seen in figure 3.1, consisting of more than 17,000 islands with 90 percent of the

FIGURE 3.1 Overlapping map images of the Indonesian archipelago and Europe

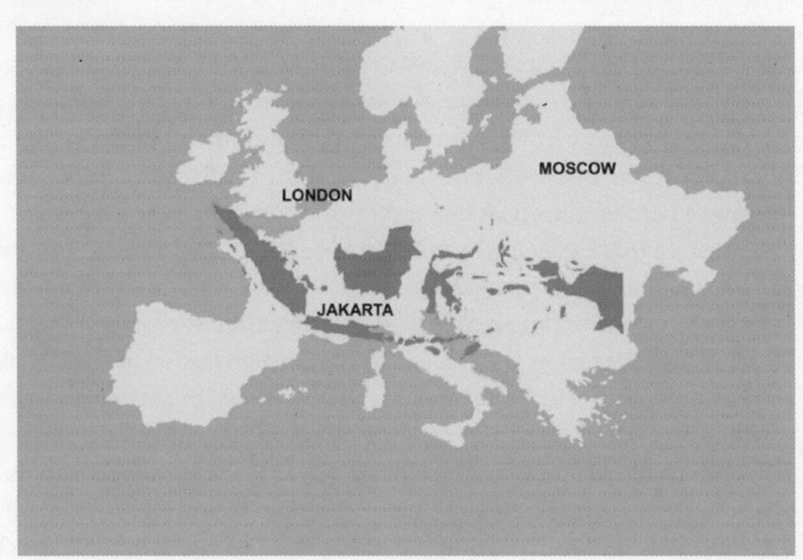

Note: The distance between the farthest western town of Indonesia and the farthest eastern part is a little more than 3,200 miles, which is equal to the land span from Ireland to the border of Kazakhstan, or longer than the distance between Seattle and New York.

population living in the five bigger islands of Java, Sumatera, Kalimantan, Sulawesi, and Papua, and, of course, the two smaller tourist islands of Bali and Lombok. In the 1970s and 1980s, consumer transactions happened only in traditional markets spread all over the country, and an effective distribution network proved to be the best approach to penetrate the market.

From Unilever's point of view:

- the development of the market through community education
- the improvement of lifestyle
- the development of SMEs to become reliable suppliers, third-party manufacturers and sales distributors by leveraging or providing trainings on technical know-how, management skills, and code of ethics

were *the most important and necessary strategic issues of Unilever Indonesia's business strategy* in developing and growing the business in Indonesia, always bearing in mind that the company is operating within *an emerging market.*

In the meantime, the SME partners benefited from the transfer and development of knowledge and skills, and were becoming substantial businesses in their own right, created by Unilever's long-term growth and continuing success. These businesses employ large numbers of Indonesians directly in their own businesses, and indirectly through the suppliers of their required materials and services. The community at large has benefited directly from the constant stream of high-quality affordable products that have improved their wellbeing and general standard of living, and indirectly from increased business and employment opportunities.

When Unilever Indonesia decided to invest in these small companies, encouraging the development of business partners which in turn created job opportunities and wealth creation, the chic term of corporate responsibility was not yet in existence to describe the cause. These activities were exactly on par with CSR, considering the wider economic and social linkage and the impact of their activities. Unilever's spectacular success of sustainable growth and profit is a stunning endorsement of the ongoing benefits of including CSR activities within the businesses strategic framework from the very early stages of its development.

Focus Point

Characteristics of a good CSR program:

- Is embedded within the business operation.
- Generates sustainable benefit.
- Provides a win–win solution.
- Is impossible for a company without profit.
- Can only be sustainable if continuous capacity building and community empowerment are done, supported by the necessary infrastructure.
- Continuous improvements through monitoring, assessing, and reporting.

In the case of Unilever, developing business partners within a company's value chain—from suppliers through production up to distribution of products in the marketplace—through using operation excellence and good business ethics resulted in a "win–win" cooperation (see figure 3.3). The company has ensured its value chain sustainability and transparency, which eventually enabled the business partners to become big conglomerates adopting the Unilever system of operation (good corporate governance).

Many people consider the establishment of an effective value chain is only for the benefit of the company, and has nothing to do with community development. This statement is true if Unilever had asked international companies to invest in Indonesia as partners within the value chain. If Unilever had chosen this particular route, it would have been easier for Unilever, as convincing international companies to be partners because, in some countries, they were already partners of Unilever. But Unilever had decided to instead develop local SMEs by leveraging operational excellence, quality assurance, and business ethics. By doing this, Unilever has contributed to the wealth creation of the Indonesia community because sustained growth of business partners created steady employment opportunities.

There are several other specific examples of the success of this approach.

Brand *Pepsodent* (Unilever)

Community education programs on oral hygiene from Unilever Indonesia initiated in 1974, which continues up to now. This activity resulted in ongoing community education in oral hygiene. *Establishment of the value chain from suppliers, manufacturing, and distribution to support this program have ensured competitive edge* and consequently substantial contribution to the sustainable growth of *Pepsodent*.

Brand *Aqua* (Danone-Aqua)

The boom in the Indonesian economy as a followup of the sharp increase in petroleum prices in the 1970s, bringing with it a large number of expatriates living in or visiting Indonesia. While industrialization grew and spread rapidly, the sources of groundwater in the main urban centers became polluted. People in Indonesia soon became aware of the risks of drinking untreated or unboiled water. Only by filtering and boiling water at 100 degrees Celcius for 20 minutes could people get clean and germ-free drinking water. This was time consuming and inconvenient, especially for people in offices or during travel. Particularly, expatriates were not immune from diseases caused by the pathogens in the local water supply. For health reasons, they only consumed bottled drinking water.

Seizing the opportunity, Tirto Utomo the founder of PT Aqua Golden Mississippi, Tbk (1973) pioneered the development of bottled drinking water market in Indonesia, with their brand of mountain spring water called *Aqua*. The first consumers were the expatriates, followed by government and private offices, soon followed by people buying for home use. Before the introduction of *Aqua*, all bottled drinking water available was imported. *Aqua* was offering a solution for the problem by introducing clear, clean, and hygienic bottled drinking water.

To ensure competitive edge, and consequent sustainable growth and profit within a vast expanding market of bottled drinking water, PT Aqua Golden Mississippi had put a major focus on the following activities. In addition to offering convenience and hygiene in drinking water, the company had continuously developed, improved, and established the extended value chain through ensuring quality assurance of *Aqua* from sourcing, bottling up to distribution in the marketplace (impact on community through the value chain).

Sourcing

Changed the source of water from groundwater into natural mountain spring water (1980), the process is strictly controlled daily and secured from any contamination.

Manufacturing

Using PET food grade material for all their packaging (1985).

To ensure the purity of every drop of *Aqua*, the hydro pro system is applied. The manufacturing technology used was the in-line process system, an automatic continuous process. The process started from blowing empty bottles, filling and closing the bottles, labeling and packing in its final packaging; all done in a sterilized room (1985).[2]

Distribution

- Define the location of the 14 factories all over Indonesia to ensure cost-effective geographic distribution in Indonesia.
- Established partnership with SMEs in developing the national distribution network to cover all segments of the market.
- To cater for all types of consumers, *Aqua* is offering different types of packaging from plastic cups of 120 milliliters (4 fl. oz.), 220 milliliters (7.4 fl. oz.; ideal for school canteens and lunchboxes), and plastic disposable bottles of different sizes.
- *Aqua* is also available in glass bottles as an exclusive serving to the guests in restaurant or hotels.
- For homes, offices, schools, and so on, *Aqua* offers five-gallon bottles provided with machines to dispense the products.

In 1998, Aqua (PT Tirta Investama) had done a strategic alliance with the Danone Group, which is one of the biggest groups of companies in the world of bottled drinking water and also an expert in nutrients. After this, the latest technology of bottling water was implemented, the quality of the product and its market share have further been improved. Currently under the Danone-Aqua umbrella, *Aqua* has a million points of distribution easily accessible to all consumers in Indonesia.[3]

Currently, the market of bottled water is enormous, and many other brands have been introduced in the market, but *Aqua* is still able to sustain its market leadership position and has become the generic name of bottled drinking water in Indonesia (see figure 3.2).

Brand *Teh Botol Sosro* (PT Sinar Sosro)

It started in 1940, the Sosrodjojo family started their business in Slawi, a small city in Central Java. The product sold was branded dried tea *Teh Cap Botol* then, with limited distribution. In 1953, the Sosrodjojo family started to expand its business to Jakarta, Indonesia's capital, by doing product tasting in several markets in Jakarta; first by offering freshly brewed tea, which was considered by the consumers too hot for direct drinking, followed by sampling already brewed tea in big pans, which was difficult to transport to the markets. Finally, the product for sampling was brewed and filled in used bottles (from other products)

FIGURE 3.2 *Aqua*

Note: *Aqua* is a pioneer in bottled mineral water in Indonesia, produced for the first time by PT Aqua Golden Mississippi and launched its first 950 milliliter (32 fl. oz.) glass bottle in 1974. The strategic alliance of Aqua and Danone Group in 1998 has placed the company as the largest mineral water producer in Indonesia.

for convenience in transportation, and ready to drink at room temperature for consumers to try and taste the tea. This new way of sampling tea was very successful.

After this experience, an innovative idea surfaced in 1969; namely, to introduce bottled "ready-to-drink tea" into the market. The idea was that you can quench your thirst by drinking "ready-to-drink tea" either at room temperature or cold without the hassle of boiling water and preparing tea from tea leaves and adding sugar into it. In 1970, a bottled tea plant of PT Sinar Sosro was built, the first glass bottled ready-to-drink tea plant in Indonesia and the world.

By introducing "bottled ready-to-drink tea" (1969), without realizing it, PT Sinar Sosro had introduced the first natural and green beverage product packed in green packaging (returnable glass bottles) in Indonesia.

Since early 1990, the business has been managed by the third-generation of the Sosrodjojo family.

To ensure focus operations (that is manufacturing and distribution), the beverage business development have since been done by two companies, PT Sinar Sosro—a company producing packaged ready-to-drink tea, and PT Gunung Slamat—a company producing dried ready-to-serve tea.

After the business growth in November 2004, both PT Sinar Sosro and PT Gunung Slamat were made subsidiaries of PT Anggada Putra Rekso Mulia (Rekso Group). To ensure competitive edge, consequent sustainable growth and profit through developing and growing market share within a vast expanding branded beverage and water market, that is, juices, *Coca-Cola*, *Pepsi Cola*, coffee, milk, tea leaves, and others, PT Sinar Sosro has done the following.

In addition to offering a "healthy hygienic natural green" beverage in convenience "green" packaging, the company had continuously developed, improved, and established the extended supply chain through ensuring quality assurance from its sourcing of tea leaves, manufacturing up to distribution in the marketplace (impact on the community through the value chain).

Sourcing

To ensure economy of scale in purchasing its tea leaves, a sister company—PT Agro Pangan—has been established, with a responsibility to ascertain the supply of the same high-quality standard tea leaves and cost-effectiveness from several tea plantations in West Java.

Manufacturing

PT Sinar Sosro is the first producer of bottled ready-to-drink tea in Indonesia and the world that uses ultra high temperature (UHT) sterilization process. Product freshness, taste, and hygiene are the focus of the company in producing quality products.

Distribution

- Establish partnerships with SMEs in developing and establishing a distribution network covering all areas in Indonesia.

- Define the location of factories based on cost-effective geographic distribution in Indonesia (total of seven factories).

All this is in line with the basic value of philosophy of PT Sinar Sosro, which is caring for three aspects and environment, that is, care for quality, care for safety, care for health, and being eco-friendly.

Currently, despite the many other brands introduced in the market, *Teh Botol Sosro* continues to grow and maintain its leadership position (see figure 3.3).[4]

All these three cases are pioneers in its own field, the same development scenario had been pursued in developing the brand within an emerging market; that is, conditioning and expanding the market through community development and education, and other efforts to improve lifestyle, at the same time developing and establishing the extended supply chain from scratch in cooperation with SMEs (social impact of the value chain).

FIGURE 3.3 *Teh Botol Sosro*

Note: *Teh Botol Sosro* is the first ready-to-drink tea in Indonesia and in the world was launched for the first time in 1970 by PT Sinar Sosro in the form of returnable glass bottle. Now it is available in different packaging and innovative variants.

In summary, they all did the following:

- Identify a potential market for a certain product.
- Develop and expand the market through a continuous process of innovative marketing and CSR activities:
  - by introducing an innovative product or packaging, in the case of *Aqua* and *Teh Botol Sosro*, they both were introducing the first "natural green products" in Indonesia
  - improving lifestyle and changing habits through community education (CSR), combined with innovative and energetic marketing
  - community education and capacity building, as well as the development of new business partners (SMEs) to ensure the establishment of the necessary infrastructure from suppliers of packaging, raw materials, manufacturing, logistics, and establishing a distribution network to customers.

The rewards for the corporations are community acceptance, resulting to an expanded market supported by an effective and reliable value chain, ensuring a competitive edge.

The SMEs benefit from help to establish cost-effective businesses and the potential to share long-term sustainable growth as the indirect result of the ongoing efforts of the bigger business to promote sales and further expand the market.

Society benefits from improved lifestyle and higher living standards from the availability of affordable good-quality products (convenience, health, and safety), the provision of more efficient services, the employment opportunities, and a range of ethical, social, and environmental standards through the acceptance of codes of business principles.

How did these companies come up with a brand development strategy that provides long-term shared benefits? The answer is in the understanding of the community behavior and needs. Box 3.1 is a story of consumers in Indonesia in the 1970s, which shows how limited their knowledge was of personal hygiene and its associated products. This provided a good base for business' early engagement in community education.

Consumers in the 1970s era

As a Senior Brand Manager of Dentifrice & Shampoo for Unilever in the 1970s, the author's main responsibility was to develop the market for Unilever products to increase sales. This job allowed the author to visit many places outside the main island of Java and sometimes hidden quarters of rural areas to meet and understand the consumers.

One of those places was Pontianak, the capital city of West Kalimantan, a marshy ground in the delta of one of the longest rivers in Indonesia, the Kapuas River. In the 1970s, the river was the main transportation system, an important corridor for settlements along the river. The river also served as public bath, wash place, as well as toilet. When you drive along the Kapuas river in the morning, you would see many people doing all the morning activities at the side of the river. There were no bridges across the river, and to visit other small towns along the Kapuas river, which stretches about 300 yards wide, you would have to have a jeep. To go to city outskirts at the other side of the river was another story; both the people and the jeep had to go on a giant bamboo raft.

The best hotel available in Pontianak at that time was a simple guesthouse, with wall panels made of woven bamboo mats. Looking out of the window, which was covered halfway with plain linen, was swamp and water. You could call it exotic, if you considered the story of the undead vampire that made the name of the city, the harsh surroundings, and the lack of clean water, exotic. This urban center made a perfect spot for customer benchmark to gauge the habits, knowledge, spending style, and other customer indicators. A good understanding of this type of consumers, which made up most of the Indonesian population, was the key to gain customer acceptance and to grow the market.

Like most of the people from Kalimantan, the women there had fair skin and were beautiful; but only when they did not smile. As soon as they smiled, we could easily see that most of the young beautiful women had very bad stained teeth. Later, we found out that the water in Kalimantan has zero fluoride content and, since most of the people didn't brush the teeth with toothpaste, most of

(*continued*)

(continued)

the population had problems with tooth decay. *Pepsodent* toothpaste was nearly nonexistent in Pontianak and its surrounding area. The few people using toothpaste were using a competitor's brand *Prodent*. Although within a still very small market, *Prodent* was the market leader in Indonesia.

During one of the visits, the author met a family and, in an interview, the father confessed that they had cleaned their teeth using toothpaste and toothbrushes. This was an exceptional situation because most of the people there either never cleaned their teeth using toothbrushes and toothpaste, but instead ground bricks, or they just did not clean their teeth at all. The father showed a completely worn toothbrush, telling happily that the whole family took turns to use the same brush every day. The family hung the precious toothbrush carefully on the wall.

This story clearly shows how limited the consumers' understanding of oral hygiene was. Although this was an extreme case, similar facts were found in almost the whole population both in urban and rural areas. Without sufficient knowledge, they would never clean their mouths and teeth regularly, let alone have bought toothpaste, because they just did not see the need. It was then realized that community education on oral hygiene was very much needed to increase the usage of toothpaste, and therefore expand the market.

THE EVOLUTION OF CSR

In the past, businesses had thought, perhaps simplistically, that development of the market was the key to sustained growth. Businesses have now realized that the future is very much linked with the society, the government and the wider environment in which it does business. Although the rationale for the existence of a business is to sustain growth and profit for its shareholders, the trend is that, over time, businesses are expected to make a significant contribution to the societies.

It has been realized that no business can expect to be successful if it operates in a difficult environment, in which high unemployment, limited education and other factors result in low and restricted economic growth. So it is the need for businesses to accept greater accountability without departing from the main proposition that is generating returns to shareholders.

CSR in a developed market started when well-known industrialists, such as Andrew Carnegie, Bernard van Leer, and Joseph Rowntree, and many other wealthy businessmen endowed foundations to carry out philanthropic works. The motivation of such philanthropy is twofold: ethical and enlightened self-interest. Today, businesses around the world recognize their obligations not only to shareholders, but also to multiple stakeholders, and see that alongside their traditional role, they also have social and environmental responsibilities.

Companies based in developing countries or operating in emerging markets face different challenges. Social investment in the form of community and capacity building is necessary and is a very important part of the business strategy. This strategy will enable business to grow in a *developing market* where basic infrastructure and the capacity to build social capital are lacking (see case study 1: PT Unilever Indonesia, Tbk).

Primary stakeholders of most companies, which are their employees, customers, and communities, are influenced by history and their culture. While social norms are local, Indonesia and other developing markets are very much affected by global factors, and the developing nation's previously parochial local marketplace has become part of a much wider regional and global market.

This has significant implications. Local markets have been, and will continue to be, increasingly competitive. Unless businesses within developing markets are able to develop true competitive edge in the products and services they offer, they risk losing their position within that market. That said, the development of a global competitive edge has even more benefits, in that it will not only ensure the survival of the business, but also give it the realistic chance to exploit the opportunities afforded by meaningful access to the wider regional and global markets. In the face of this competitive environment, if a developing country is to become a positive and an encouraging environment for new investments (both local and foreign), it must meet the new global challenges. The responsibility to provide the necessary conducive community and market environment, and the means of developing global and regional competitive edge required, are not solely the governments, but that of all the stakeholders, including the business corporations, NGOs, educational institutions, and the community.

The environment for sustained growth and profit will be achieved with the universal application of good corporate governance, the incorporation of the precepts of CSR within the business strategy of each individual company, and the widespread acceptance of corporate social responsibility throughout the whole cooperating community. A logical approach to achieve significant advancement in CSR attitude within a company is to satisfy the spiritual needs and values of the employees and their families. If they are to become good corporate citizens, the right perception allows them to participate actively in CSR initiatives and activities and this will help guarantee and retain highly capable human resources (see figure 3.4).

FIGURE 3.4 The framework

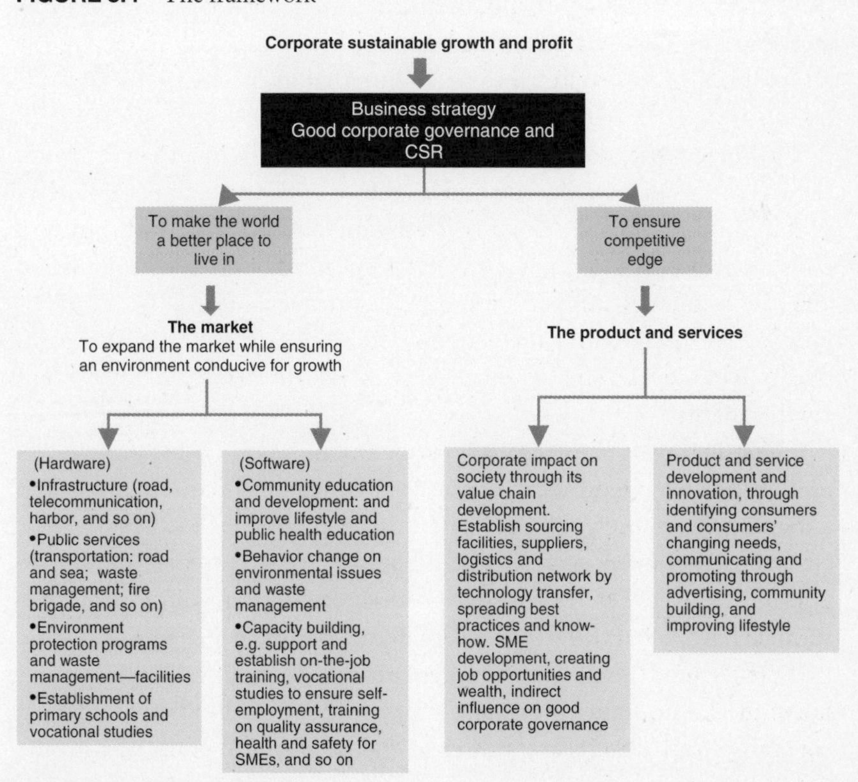

The Market

In the past expanding the market (within an emerging market), involved community building, education, improving lifestyle through the introduction of products to meet basic consumer needs; for example, soap, toothpaste, shampoo, teabags, ready-to-drink tea, and instant noodles. In those days, a company could only grow when it had a product the people needed, and when it was able to produce and distribute the product in an efficient way.

Nowadays, if the basic market is to be expanded and experience sustained ongoing growth, it must be conditioned to ensure that it is a favorable environment to attract further foreign direct investment. As a result of globalization, there is a greater awareness of all the characteristics and attributes of a much wider range of available products, including those incorporating the most advanced technology.

Still, the same basic rules of business apply. Unless the consumer can afford the product offered, the corporation will not achieve sustainable growth. The only thing that the corporation can do to ensure the affordability of its product is to make sure that its production and distribution are carried out in the most efficient way.

However, affordability is often a function of macro and micro aspects of the wider economy and therefore outside the direct influence of the company. If the economy is growing and thriving, with plentiful employment opportunities, a greater proportion of the total population will find a greater number and range of products affordable. Employment opportunities are largely a function of investment, so it is imperative that a developing country is seen as an attractive investment environment.

High unemployment and still struggling microeconomic growth will reduce the average purchasing power of consumers, particularly within a developing country. At the same time, the effect of increasing labor costs has not been offset by the benefits of vocational training to improve skills levels. As a result, production efficiency and quality assurance have fallen below the acceptable standard.

Furthermore, weak law enforcement that leads to failure to implement good corporate governance more widely could inflict a high cost economy. Low purchasing power of the population, coupled with the high cost of operations, including the high distribution cost, could drastically reduce the attractiveness of a developing country as a local market, let alone to compete successfully as a sourcing country for the regional or global markets.

The Product or Services

The evolving paradigm of business within the global competitive environment combined with the change in employees' values and active participation of NGOs concerning humanitarian and environmental issues demand smart and fresh business approaches. Identifying and anticipating consumer's changing needs through new means and tools are very important, besides ensuring the positive impact a corporation has on society.

If companies are to ensure a competitive edge required to survive and operate in a regional and global market, implementing the principles of CSR in their vision and mission, as well as in their business strategy, are of vital importance. Over the past few years, we have seen more and more organizations take up the idea, and there is enough evidence to confirm that this initiative will ensure "competitive edge" and "sustainable growth."

This new understanding of operating a sustainable business following the principles of CSR, that is, profit, people, and planet, is supported and encouraged by the availability of international guidelines such as the Global Reporting Initiative (GRI), the Global Compact, the OECD Guidelines for Multinational Companies on behavior patterns to be followed when operating globally including in the developing and emerging markets, and the formation of the ISO 26000 involving most countries globally through its mirror committees.

TYPES OF CSR

Previously, all CSR activities were seen as the full responsibility of each corporation. Currently, it is more appropriate to divide CSR activities into three categories, not including corporate philanthropy (see table 3.1).

1. The first category is where the choice of activities and the partners involved are the full responsibility of the corporations. The result gives

TABLE 3.1 The types of CSR and the expected benefits to the company and the community

Type of CSR	Benefit to community	Benefit to company	Budget
CSR linked with the extended supply chain	Creating employment and wealth Best practices in operation Good, profitable SMEs Indirect influence on sustainable good corporate governance	Good reliable sourcing facilities Internal high machine efficiency Operational excellence and competitive edge	Part of operational cost
CSR linked with market development and expansion	Sustainable Behavior training and change Improved lifestyle	Sustainable Increase imaged and usership Brand equity Competitive edge	Part of profit Unconventional media budget (Part of advertising and promotional budget)
CSR to improve lifestyle and ensure a market and environment conducive for growth	Sustainable benefit Community capacity building and empowerment Behavior change on environmental issues and waste management Community education, public health, infrastructure	Long-term sustainable benefit Corporate image and community goodwill Competitive edge	Allocated budget for CSR could be: part of profit + advertising budget + operational cost
Corporate philanthropy	*Ad hoc* direct benefit	No sustainable benefit	Part of profit

direct benefit to both the corporations and the community and society. Activities that fall into this category are: CSR linked with market development.

The first brands entering the market and developing or expanding the market through community education, community building and development, and also improving the community lifestyle, will usually sustain their market leadership, provided that the brand maintains ongoing innovations, creative and energetic marketing, and other combined CSR activities to support the brand.

Sustainable marketing works:

- to sustain the competitive edge of the brand and its supporting operational processes (extended supply chain) through continuous innovation to ensure the brand stays at least one step ahead of its competitor.
- to continue to develop and expand the market through educating and developing the community and improving the lifestyle of consumers through appropriate CSR activities, combined with innovative and energetic marketing.
- to ensure an environment conducive for growth through care for community wellbeing and environment safety.

CSR is linked with the extended supply chain (and impact on the community through the value chain). *Extended supply chain* is a term used to describe parties involved in the creation of a product, from sourcing through production to the marketplace (see figure 3.5). In implementing CSR within its extended supply chain, a company establishes and develops SMEs through guidance and training in best practices and good governance, using technology and know-how as a means of outsourcing production, distribution of sales or logistic facilities, and sourcing facilities for raw and packing materials.

The development of these mutually beneficial partnerships not only results in increased employment opportunities and availability, but contributes substantially to *the creation of wealth* through the advancement of well-run and profitable enterprises. The company further benefits from its ability to increase its competitive edge by concentrating its own efforts to achieve higher levels of internal efficiency and operational excellence.

FIGURE 3.5 Value chain development from suppliers, production process, and distribution to consumers

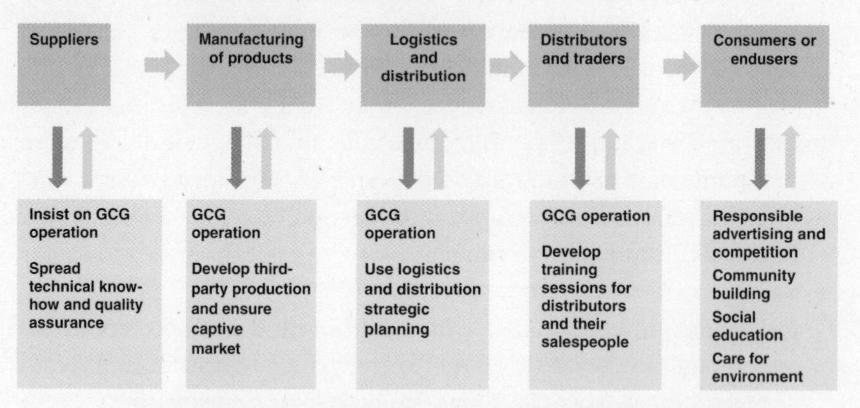

Corporate impact on society (SME) through value chain development

There are examples of most successful corporations that are implementing this type of CSR; making a valuable impact on the community through the value chain. Readers can learn the specific implementation from companies such as Unilever Indonesia, PURA Group, Astra International, and many others in the case studies described in detail in part II of this book.

Focus Point

Investment in long-term supplier development and partnerships with SMEs creates reliable supply of products and services, giving the company an efficient supply chain that ensures competitiveness and long-term sustainable business for SMEs.

2. *The second category*, which sometimes is also called community relations, is where the choice of the CSR activity depends on the needs of the community or society in which the corporation is operating, and where it gives sustainable benefit directly to the community and society and indirect benefit to the corporation, such as community acceptance, image building, and reputation risk mitigation (see box 3.2). These types of activities include:

- community and capacity building through:
  - supporting basic education and establishing vocational studies and on-the-job training programs to ensure self-employment during periods of scarce job availability (see Astra International case study in part II for a specific example)
  - guiding and training SME management in simple accounting systems, best practices, good governance, and quality assurance. Corporations can also use technology expertise to create simple affordable equipments to help SMEs to operate more efficiently (see the PURA Group case study in part II)
- environmental safety, community wellbeing, and public health and safety:
  - the support of behavioral changes or changes of mindset can contribute toward a far-reaching impact on the wellbeing of the wider society. Several well-accepted programs, such as "My Clean, Healthy, and Prosperous Market" launched by PT Bank Danamon, Tbk (see PT Bank Danamon, Tbk case study) or Unilever's environmental care program in Surabaya, are setting a good example in their effort to encourage waste management and in stopping the pollution of rivers and canals to avoid floods. To ensure sustainability, a new and better living concept has to become an everyday habit and, consequently, continuous training, facilities and infrastructure should be provided
- volunteering:
  - a type of contribution to community projects by employees during their free time, with the support of the business or employer. Employee participation in CSR community projects will increase pride and loyalty, and is an effective means to retain capable human resources

Community Relations

For CSR of the second category, a project from Unilever in Indonesia can be used as an illustration. The year was 1995, when the government of Indonesia announced a new regulation that allowed only iodized salt sold in the market. Only *PN Garam*, a state-owned

(*continued*)

(continued)

salt enterprise was producing iodized salt, while much of the salt consumed and sold was produced by salt farmers in Madura, a small island north of East Java, and also from salt fields along the north coast of East Java. The farmers used only the simple traditional method of evaporating saltwater by flowing seawater into large drying salt fields. With the new regulation, no farmer could sell the harvested raw salt in the market because they did not have the needed facilities to process and iodize their salt. In the duration of three months, all noniodized salt would be taken out from market.

With a factory operating in the industrial estate within the same region, just a few miles away from the salt fields, Unilever believed that the company should do something and give back to the community to ensure sustainable operation in that particular area. A prompt decision was made, and Unilever Indonesia had initiated the building of a salt iodization facility for the local farmers with capital investment of IDR1 billion (about $500,000). Two engineers were sent to the premises to help build and run the facility for two years. At the same time, salt farmers received capability training to provide them with the necessary skills to operate the facility independently, so after two years the facility was fully transferred to the salt farmers association. All these activities were supported by the provincial governor and local government.

Immediate benefits after the implementation and operation of the facility were realized only by salt farmers and the community, with no direct benefit for Unilever. In 1998, however, when the Asian financial crisis hit the country, and most factories located in the same area faced difficulties dealing with workers' protests against business actions taken by the companies, Unilever was able to operate without any disturbances. Unilever has gained trust and confidence from the local community, which proved to be very important.

The indirect benefit to the companies of all these activities is that the generation of jobs in the community and in the surrounding areas will create wealth, which will result in positive microeconomic development, and contribute to higher purchasing power. The benefits to the community can be further increased by companies operating in the same area joining forces to work together to further develop the area.

3. The third category is corporate philanthropy, which is often seen as the traditional CSR activity. This involves the direct contribution to charity in the form of cash grants, donations, or services in kind at the time of national disasters, for example. While it does provide some short-term relief, it is often misguided, is of limited long-term benefit, and can result in unwanted problems. One popular philanthropy program that gives only one-off benefit with no sustainability is the *Sunatan Massal* or free mass circumcision surgery, which is done by many companies in Indonesia because circumcision is a must for Moslem's young boys (aged five-to-12 years), and many families cannot afford to do it, so it is considered a good CSR activity to help families and the community.

Implementation of CSR in this category is usually based on *ad hoc* needs, and is difficult to monitor. It is usually not possible to measure the sustainability of the results of this type of CSR and worse, the funds are often misused to cover deviations from the GCG norm through bribery, fraud, and other malfeasance.

NOTES

1. Doing Well by Doing Good—75 years Hindustan Unilever Limited, special issue.
2. http://www.aqua.com/aqua_v3/ina/danoneaqua_aquacare_detail.php?p_id=1&act=next.
3. http://www.aqua.com/aqua_v3/ina/danoneaqua_about.php.
4. http://www.sosro.com/sejarahbisnis.php and http://www.sosro.com/profil-perusahaan.php.

4

HOW TO OPTIMIZE CSR

While people may still be troubled by both the unclear definition of corporate social responsibility (CSR) and by the fact that it has frequently been viewed as something separate from a company's primary business, all indications are that there is greater clarity about the prerequisites for success.

The following are prominent examples that suggest that the best chance for sustained growth and profit comes when a company has embedded CSR into its strategic business strategy and continues to develop the market to ensure that consumers understand and accept the benefits of using its products or brands or services, and that they also can afford to buy them.

- PT Unilever Indonesia, with its brand *Pepsodent*, started market development through its program of oral hygiene education since 1975, supported by the development and establishment of an effective value chain that continues up to the present time (see the PT Unilever Indonesia case study in part II).
- PT Bank Danamon, with its innovative community banking program, called Self-Employed Mass Market (SEMM) or generally known as Danamon Simpan Pinjam (DSP), has started market development through a program of My Clean, Healthy, and Properous Market (see the PT Bank Danamon case study in part II).

Before starting any CSR activities, a business should ensure it has sound internal corporate governance and define which sustainable CSR activities it should promote and develop. This has to be done even at the earliest stages of a new investment. It may be to develop the market and improve lifestyle through community building, or by maintaining its image in conserving the environment, or to have an impact on the community through its value chain, or anything else that would increase the mutual benefits of the company and the community. In addition to the needs of the business corporation itself, a survey identifying the community needs and conditions within the area in which the business is operating is very important.

It is a prerequisite that businesses should realize that for *successful and sustainable implementation* of community CSR activities, the means or infrastructure should be provided by the corporations; for example, development of community capability, using technical know-how, and sponsoring vocational studies. Continuous monitoring and training have to be done to ensure the community's newly employed and learned values have to become a habit. It needs time and patience.

Particularly, for nonconsumer product businesses, the traditional view of sustainability is that a business should not use up the natural resources it depends on, but ensures ongoing supply through a continuous program of maintenance. For example, a timber company would replant trees to replace what is harvested, or a paper manufacturer would not only harvest, but also develop its own pulpwood (pine, eucalyptus, or other softwood trees) plantation. This kind of sustainable harvesting system would not only serve the company's own interests, by maintaining its own source of supply, but also serve to preserve the environment and avoid floods, landslides, and other natural disasters. In the case of the mining industry, when all the mineral resources are exhausted, the mining company could ensure sustainability of community income in the area surrounding the mining sites, by providing income-generating advice and funding the training and development of other skills for the community. The nature of this support would depend on the need and the availability of resources in the area and could include such projects as farming, fishery or cattle and horses breeding (see case study 5: PT Indo Tambangraya Megah, Tbk). With the increasing awareness of social obligations, mining companies have no choice but to maintain community acceptance to ensure sustainable growth and profit. It is very important to remember that any CSR activity involving a wider community should be in the form of a joint effort of the business corporation with the community, local government, and nongovernmental organizations (NGOs), and supported by expert or educational organisations. NGOs can help monitor, control, and assist in establishing the industry standards while educational institutions can give assistance on their expertise or knowledge not available within the company, such as with the example of Unilever's cooperation with the dentist association. The main responsibility of the implementation still lies within the business corporation. The company should lead the activities, holds responsibility of the implementation and reporting, leverage, and insist on good corporate governance. This involves employees actively in solving CSR issues and in the implementation of CSR activities.

The possibility of joint CSR budgets from several corporations could represent a substantial potential source of public revenue, and could be used as a base for increasing employment opportunities and further wealth creation. In a developing country, several corporations' collaborative CSR activities could contribute substantially to the development and acceleration of microeconomic sustainable growth, through using good governance, value change, best practices, and technological know-how. There are several illustrations of the synergetic flow-on benefits of joint CSR efforts.

HUMAN RESOURCES

Vocational Studies

One strategic area where CSR can contribute significantly is human resources, with which a company can develop programs that promote employment creation through support of basic educational studies and capacity building. This can be done by establishing vocational programs to ensure either self-employment or the availability of trained and skilled labor for external and local needs, for example, nurses, blue-collar laborers, and cooks, or even the possibility of benefiting from the wider experiences, expertise, and remittances from workers trained to work outside the country.

A good example is the CSR program of Astra International, one of the biggest diversified business groups in Indonesia. The company is committed to provide training and assistance to school dropouts, retirees, and unproductive citizens, empowering them in the economic structure. The detailed vision and implementation of the initiative can be found in the case study on PT Astra International, Tbk in part II.

Tertiary Education

It is realized that a developing country such as Indonesia with a large population (around 230 million in 2009), needs future leaders who have gone through tertiary education with high integrity and good value. With the trend of increasing cost for tertiary education globally, including Indonesia, only people with money can enter universities. The following is a case that can be used as a model of how to maximize a joint CSR budget effectively.

Putra Sampoerna has established "The Sampoerna Foundation" (SF) concentrating on "Creating access, affordability, and opportunity for future young leaders", and also to give everybody in Indonesia the same opportunity to enter tertiary study both in Indonesia and abroad. In establishing the foundation, Sampoerna has built the necessary infrastructure to serve as a base for partners and beneficiaries to contribute to and participate in without starting from scratch in defining which educational CSR activities need to be done.

SF is developing the services shown in figure 4.1 to be shared with partners and other beneficiaries.

In maximizing coverage and effectiveness, the Sampoerna Foundation is structured into three different parts (see figure 4.2):

- the back office: in which all operational activities of the foundation are done, that is, the enablers and the core services
- the middle office: in which CSR and philanthropic programs are implemented both SF's own and joint programs with donors or beneficiaries from other corporations
- the front office: sales and marketing by SF

FIGURE 4.1 SF's services

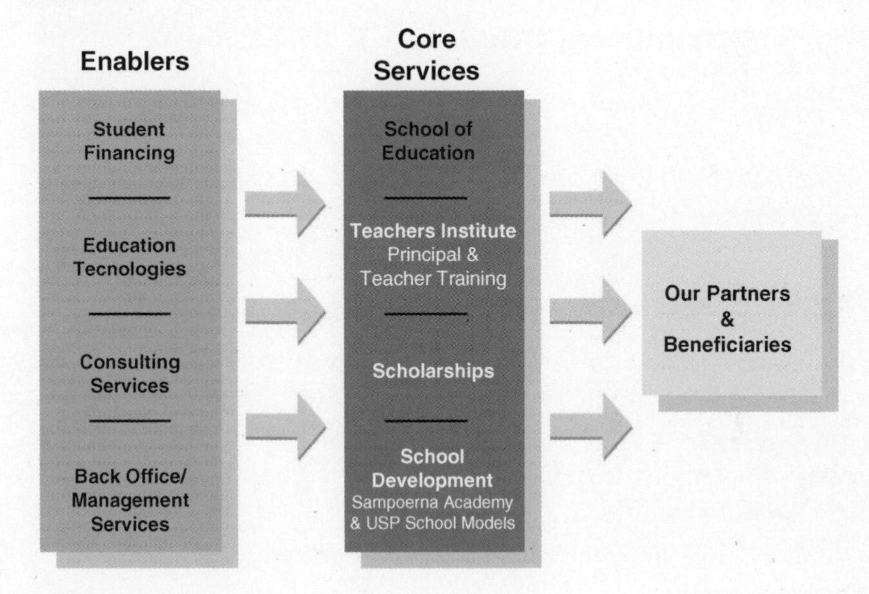

The ideal situation would be that the SF focuses more on running the back office and the front office effectively, thereby maximizing the implementation of joint educational CSR activities from partners and beneficiaries.

Given that integrity and values are important requirements to be scholars of the SF, alumni of the Sampoerna scholars have to do four social programs a year, monitored by the foundation, as described in the following.

Sponsor Services—Sponsor Benefit

Benchmarking

- Measuring progress and achievement
- Whole-school "Key Quality Indicators" (the school's goals for learning and performance, teaching and learning effectiveness, organizational effectiveness)

FIGURE 4.2 Division of functions at SF

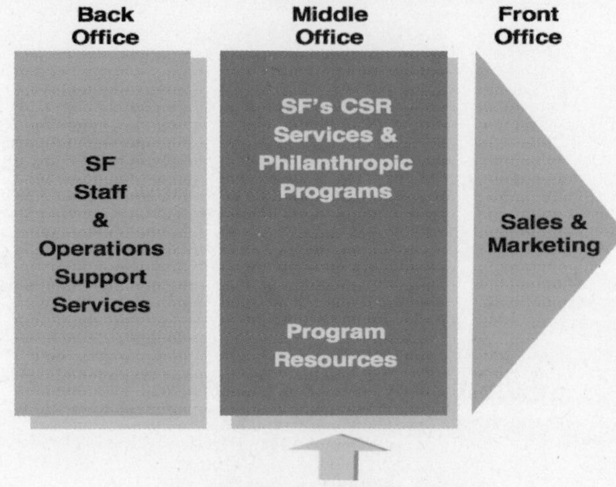

Note: Supported by the infrastructure of the back office and the front office, any partner or beneficiary can contribute and participate in its choice of educational programs available. Currently, most of the joint CSR budget is still contributed by Putra Sampoerna.

ISO 9001

- Accountability for use of funds
- Quality management system
- Sponsor feedback

Optimization

- Making the most of available resources and greater coverage of funding

Result As figure 4.3 shows, this established a pathway to creating young leaders:

- scholarship for students:– 30,000+ over six years
- scholarships for overseas MBA students

FIGURE 4.3 Pathway to creating young leaders

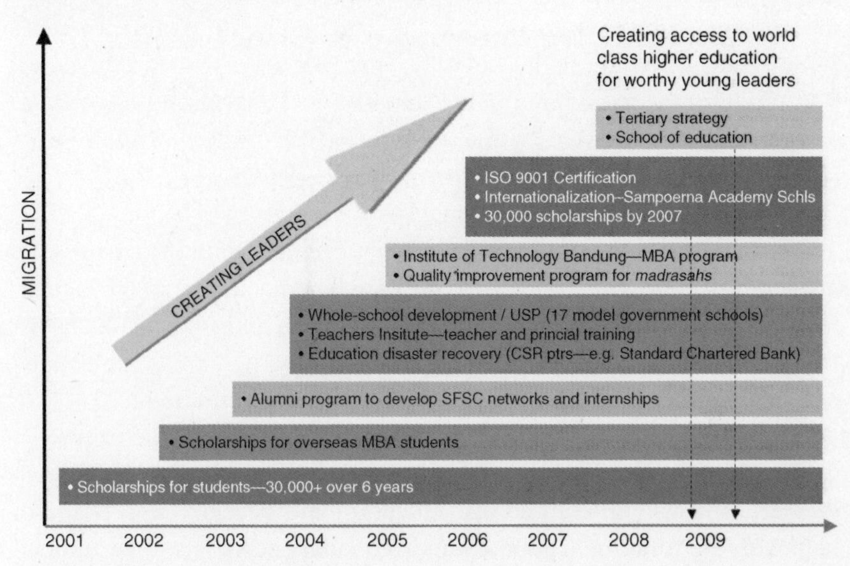

Source: CSR & Philanthropic Education Services (June 2009) by Ron Perkinson, President, SF.

- alumni program to develop Sampoerna Foundation Scholars Club networks and internships
- whole-school development and United Schools Program (17 model government schools)
- Teachers Institute: teacher and principal training
- education disaster recovery (CSR partners—for example, Standard Chartered Bank)
- Institute of Technology Bandung: MBA program
- quality improvement program for *madrasahs* (Islamic schools)
- ISO 9001 certification
- internationalization: Sampoerna Academy and *madrasahs*
- tertiary strategy
- school of education

Technological Know-How Leverage

SMEs traditionally suffer as a result of their limited management skills. For example, when they go to the bank to obtain capital, they might not make a good impression because of their lack of knowledge of accounting and their presentation of their investments and repayment plans. Often, they also don't know where or how to market their products, or don't know how to ensure quality assurance. Their growth prospects are substantially boosted by the adoption of the lessons of their partners' experience and the absorption of their expertise.

Joint CSR budgets could become the source of funds to ensure the availability of good infrastructure, good education, good health facilities, well-trained human resources or labor, and care for the environment. This will help accelerate the development of a favorable market environment and, as a consequence, the attractiveness of a developing country like Indonesia to both local and foreign investors.

These joint budgets could also complement the state national budget, especially in community education and development. It could help with the funding of primary and secondary schools and vocational studies, which are the foundation for creating self-employment, as well as helping to satisfy the basic needs of a skilled and efficient workforce. Corporations could offer on-the-job training for graduates from vocational studies. A better educated workforce could justify increased earnings and increase purchasing power to fuel further market growth.

Tax incentives could be considered to encourage corporations to invest in CSR activities, which, while giving direct benefit to the community, would result in indirect benefit to the companies themselves.

To ensure optimum national impact, it might be appropriate that all CSR activities giving direct impact on society be coordinated by a coordinating board consisting of representatives of all stakeholders; including, central and regional government, the corporations, NGOs, educational institutions, the media, and so on. The responsibility for the operational handling of the CSR activity should remain with the business or corporation owning the activity. Reporting should follow accepted business accounting principles. To ensure transparency and good corporate governance, the reporting should be monitored and controlled and openly reported to the public in line with the Global Reporting Initiatives (GRI)[1].

The sum benefits of joint CSR activities based on the different needs of each individual corporation pursuing ongoing growth and development through the common commitment to the development of a sustainable conducive market environment and community wealth creation, coordinated and supported by the government, will make a massive contribution to ensure the successful achievement of the national welfare, and in a broader view, the Millennium Development Goals (MDG).[2]

NOTES

1. For more, see chapter 6.
2. In September 2005, 189 nations, including the government of Indonesia, signed the United Nations Millennium Declarations. In signing the Declaration, Indonesia affirmed is commitment to reduce poverty, improve health and education, and promote peace, human rights, gender equality, and environmental sustainability. The Millennium Development Goals (MDGs) emerged out of the declaration, setting specific measurable targets to be achieved by 2015.

5

WHY AND WHEN TO APPLY CSR ACTIVITIES

There are several reasons behind the need to define CSR as business strategy. Aside from major regional and global competition, which would certainly have a significant impact on business, corporations must also consider other external challenges, as well as internal needs, to survive.

EXTERNAL CHALLENGES

Economic globalization creates competition, opportunity, and multiple challenges. At the same time, unprecedented advancement in technology, especially in the area of information and communications, results in more demanding consumers.

The logical consequence of a developing market—an unstable environment with high unemployment and limited education—is consumers with low purchasing power.

Among the external pressures include arguments from organizations and forums such as Oxfam and World Social Forum pinpoint issues that corporations have to deal with. These include:

- corporate capitalism that creates consumerism for underprivileged consumers; advertising that creates unnecessary needs;

multinational companies, as well as big local corporations, kill smaller local companies

- environmental issues ranging from industry litter and industrial pollution, problems associated with packaging or detergent that spoils the land and water and mining that ruins the soil

INTERNAL NEEDS

- To promote an environment that is conducive for sustainable growth.
- The vision and commitment of responsible corporations and their employees to contribute to sustainable micro economic development, by sharing best practices and involving the community in the company's value chain
- The company's beliefs that it should offer, not only excellent products and services, but also contribute to make the world a better place to live in
- To satisfy spiritual values of the employees

The initial stage for implementation of corporate social responsibility (CSR) requires companies to identify programs that fall into the first category of CSR: activities giving direct benefits to both the company and the community. In this win–win situation, a company is able to expand its market and promote a good image when the said activities successfully develop the community; for example, in improving consumer lifestyle or promoting health consciousness. In the case of the mining industry, it is especially important to improve public image and gain community goodwill and acceptance through meaningful and strategic CSR programs.

Unlike consumer products, in which advertising could give a competitive edge to a product, mining industry products can only be differentiated from their good or strategic CSR activities. Modern communications technology and the power of the media can either benefit or ruin the business, because information can easily reach every corner of the world. A multinational company operating in different countries should not ignore this situation, and should therefore ensure that it operates responsibly toward its people, community, and environment in all countries in which it operates.

For CSR projects that are linked to the extended value chain, a company can work out its business strategy by defining the focus of operation and deciding on processes that will be outsourced to third parties as an early step for developing small and medium-sized enterprises (SMEs) through several different services: training, technical know-how, quality assurance, or influencing good governance; for example, creating well-managed and profitable SMEs as a result of this endeavor generates employment opportunities within the community, enabling them to get involved in the formal economy, and consequently create wealth. This in turn would benefit the company by enabling it to operate in a more efficient manner and to concentrate on processes that would enhance its competitive edge. Building partnerships with SMEs would include sharing long-term business plans.

Having established the embedded principles of CSR within a company's business strategy, thereby attaining sustainable success, a company can expand further its CSR activities to increase the continuing wellbeing of the community. Investment in CSR programs, such as contributing to national education, providing vocational training, supporting the building of infrastructure, or working on waste and water management and other environmental issues, some may not provide direct tangible benefit to the company, but all these activities will result in risk mitigation, increased brand value, creating goodwill, improving staff efficiency, and morale, and more importantly will accelerate microeconomic growth, and consequently ensure an ongoing conducive environment for the company to operate and grow.

PRACTICAL ADVICE IN IMPLEMENTING CSR

Careful consideration is clearly required before starting any CSR activity. The chosen activity should also start within the closest boundary of influence; that is, the core business operation:

- Ensure a top-down commitment of the CEO and board and secure good corporate governance in the organization is in place within the core business operation.
- Initiate "focus group discussions" among stakeholders, that is, all groups and individuals connected to the business (business partners,

government and regulators, consumers, society and community, media, nongovernmental organizations (NGOs)), because these people can identify the weak and strong points, failures and successes of the corporations' social, environmental, and economic aspects (a SWOT analysis of the business).

- The basic strategy for CSR has to be defined at the executive level, and the company's subsidiaries can adopt this strategy according to the local environment. However, the accountability of the implementation can stay with the divisions and subsidiaries, which can modify the CSR programs in line with the local society's needs and the capacity and capability of the business to employ know-how and to ensure the success of the programs.
- Involve and empower the community in market development and image building programs (community education, improvement of quality of life, changing habits) and in the value chain development of the business (transfer of technology and systems, quality assurance, and sharing good corporate governance with all partners), resulting in expanded employment opportunities, thereby creating wealth and contributing toward sustainable microeconomic development.
- In the case of CSR programs, such as the environmental program, which gives indirect tangible benefit to the company, it is recommended *to establish a CSR unit within the company completely separate from the operation*, but ensure that all the plans, budget, control, and reporting follow the GCG principles and are incorporated into the "long-term plan" of the business. Monitoring needs to be done through safety and environment surveys to measure community acceptance and the corporate image, and to provide reasonable assurance to stakeholders.
- Define community and capacity-building programs that cover the support of basic education and vocational studies, environmental safety, community wellbeing, and public health and safety. Examples are helping the community establish and train a full-fledged fire brigade and infrastructure.
- Care for the environment through supporting behavior change or change of mindset (for example, stop polluting rivers and canals with rubbish to avoid floods), or to recondition soil in the case of mining, waste and water management, reduced waste, and so on.

- Form alliances with NGOs, central or regional government, educational institutions, and media.
- Employ business best practices—product quality assurance, good corporate governance and code of ethics, practice anticorruption, operational excellence, care for environment, and so on—with all partners.
- Collaborate with other corporations in promoting good corporate governance and CSR activities through jointly sponsoring seminars and workshops, as well as tutoring and mentoring the implementation process.
- To ensure continuous monitoring of and assessing the program's implementation, to learn from mistakes and achievements in order to move forward and get the most benefit from becoming more socially aware and responsible, and to issue sustainable reporting following the Global Reporting Initiative standard.[1]
- Reporting effective CSR activities should be exposed clearly in the media to encourage other institutions and companies to follow suit. Although public concern about the social and environmental responsibility of corporations has increased considerably, awareness of responsible practices is low and people want to know more.

To optimize the highest positive impact on the country or microeconomy development (to create employment, wealth, care for the community, education or capacity building, and care for the environment). Collaboration of different corporations implementing similar CSR activities are encouraged. The following is an insightful quote from "The Enterprise of the Future, IBM Global CEO Study (2004, 2006, 2008)," which is in line with the author's experiences outlined previously:

> Many CEOs are already moving beyond doing good and are growing their businesses by being more socially responsible. Here are some ways the Enterprise of the Future approaches CSR even more holistically:
>
> —*Understands CSR expectations*
>
> Too many companies find themselves relying on assumptions about what CSR means to their customers. Only one-quarter of

the companies surveyed in a recent CSR study said they understood customers concerns well. But the Enterprise of the Future knows what its customers expect. It uses facts and direct customer input as the basis for its decisions.

—Informs but does not overwhelm

The Enterprise of the Future is transparent, but unobtrusive. It finds creative ways to provide relevant information, such as codes on packaging that allow interested customers to look up details—sourcing information, potential environmental impact and recycling instructions—while in the store or later at home.

—Starts with green

Given the price of oil and rising concern over carbon emissions, energy efficiency is critical for businesses as well as our planet. The Enterprise of the Future often begins its CSR changes with environmental initiatives. Through these efforts, it learns more about how to effectively collaborate on issues that affect us all.

—Involves NGOs as part of the solution

Instead of being wary of activist groups—or simply reporting data to them—the Enterprise of the Future collaborates with them. For instance, it might enlist NGOs to help monitor and inspect facilities or to assist in establishing industry standards.

—Makes work part of making the world a better place

Prospective and existing employees want to work for ethical, socially responsible organizations. But the Enterprise of the Future understands that workers also want to be actively involved in solving CSR issues. Its initiatives rally employees together in a cause that literally makes the world a better place.

NOTE

1. See chapter 6.

6

MANAGING, MONITORING, AND REPORTING

With corporate social responsibility (CSR) activities becoming an important part of business strategy, it is necessary to monitor and control the budget allocated to CSR, and measure the result accordingly. Good corporate governance requires that reporting of CSR activities should not only comply with accepted business accounting principles, but it must also be objectively monitored and be transparently available to the public.

Only CSR activities, directly linked with the establishment and operation of the value chain and with a clear direct benefit for both the company and the community, monitoring could be done through the company's system and the reporting of results are included in the company's financial reporting system (including internal and external audit and risk management control). While for all other CSR activities, a separate social accounting report is required (supported by social auditing done by a fully qualified independent social auditor), and usually is included in the annual report as appendices.

Social accounting is a method to define the value of the impact of businesses operations on society. This could include the impacts on the environment; waste, air pollution, and water management; the effect on the local community; and the cooperation and communication with its stakeholders and others.[1]

Social auditing is a process in which an organization can account for its social performance, and report on and improve that performance. It assesses the social impact and ethical behavior of an organization in relation to its aims and those of its stakeholders.[2]

According the International CSR Standards and Norms, prepared by the Council for Better Corporate Citizenship (CBCC), an organization established with the full backing of the Japan Business Federation that has thoroughly systemized CSR activities, CSR is becoming an increasingly prominent and important issue for the following reasons.[3]

- *Concerns over globalization of corporate activities:* Some non-governmental organizations (NGOs) and developing countries claim that globalization is widening the gap between the rich and poor and hurting the global environment.
- *Greater social awareness among consumers:* Consumers are pushing harder for companies to protect the environment, respect human rights, and abide by fair labor standards.
- *Corporate behavior judged by investors—socially responsible investment (SRI):* When deciding where to place their money, investors are increasingly making choices after considering companies' commitment to social responsibilities.
- *A more conscious workforce:* People looking for employment are increasingly likely to examine a company's CSR record before applying.
- *Legislative changes, primarily in Europe:* Governments have enacted legislation promoting greater CSR and SRI.

Different efforts are also being initiated by world organizations, forums, and NGOs to define a global standard for CSR boundaries, monitoring, measurement, and evaluation. Although reporting is not yet mandatory, it is a very important tool to help the company to convey to the public how effectively their business is being managed, and at the same time alert the employees to the company's performance. CSR reports also give an indication to the managers and employees that the company takes its social values policies seriously. The global professional accountant body, the Association of Chartered Certified Accountants also promotes social accounting within its own profession to support business in dealing with resolution of sustainability issues and reporting.

There are currently quite a few international CSR reporting systems and guidelines available as a result of the growing appreciation of reporting of companies' nonfinancial performance. The following are the various CSR standards and norms as compiled by the CBCC in its effort of formulating standardizing international CSR standards.

GRI Guidelines

The Global Reporting Initiative (GRI) is a large, independent multistakeholder network, launched in 1997 as a joint project of the American NGO Coalition for Environmentally Responsible Economies (CERES) and the United Nations Environment Program (UNEP). GRI became independent in 2002.

GRI issues a sustainability reporting framework, which has been adopted by more than 1,500 companies in 60 countries. It has become the *de facto* global standard for reporting. The GRI Guidelines offer principles and indicators to measure a company's economic, environmental, and social performance, as well as standards for the content of corporate sustainability reports.[4]

Its mission is to develop and disseminate globally applicable Sustainability Reporting Guidelines. These guidelines are for voluntary use by organizations for reporting on the economic, environment, and social dimensions of their activities, products, and services.

The GRI incorporates the active participation of representatives from business, accountancy, investment, environmental, human rights, research, and labor organizations from around the world.

GRI is an official collaborating center of UNEP and works in cooperation with the UN Secretary-General's Global Compact.

Green Paper 366

Issued by the Commission of the European Communities, Green Paper 366 promotes the European framework for CSR. After taking into consideration the opinions of many sectors on the Green Paper, the commission issued a White Paper indicating strategies needed to promote CSR practices at the European Union level.[5]

Caux Round Table Principles for Business

Caux Round Table is an international network of business leaders promoting moral capitalism and responsible government. The first set of Principles for Business was drawn up through collaboration among Japanese, American, and European business leaders.[6]

The Global Compact

The Global Compact is a strategic policy initiative issued by the UN for businesses worldwide to perform their operations in a sustainable and socially responsible manner, and also to report the progress of members' implementation. Businesses are encouraged to align their operations and strategies with 10 universally accepted principles on human rights, labor standards, environment, and anticorruption. By participating in the UN Global Compact, companies and organizations are able to gain access to rich knowledge and expertise, management tools, and resources offered by the UN properties.[7]

The two main objectives of The UN Global Compact are:

- Mainstream the 10 principles in business activities around the world.
- Catalyze actions in support of broader UN goals, including the Millennium Development Goals (MDGs).

OECD Guidelines for Multinational Enterprises

Organization for Economic Cooperation and Development (OECD) provides guidelines recommended by participating governments, suggesting certain behavior patterns to be followed by multinational corporations. Aside from its 30 member countries, the organization has currently broadened its focus to include other countries as well as developing and emerging economies.[8]

SA 8000

These are standards on human rights and ethical behavior, drawn up by Social Accountability International (SAI), an American NGO, to eliminate unfair and inhumane labor practices. The SA 8000 is based on the International Labor Organization (ILO) conventions, the Universal Declaration of Human Rights, and the UN Convention on the Rights of the Child.[9]

AA 1000

The AA 1000 Series are standards developed by the Institute for Social and Ethical Accountability or Accountability's, a global nonprofit organization with bases in London, Beijing, Geneva, Sao Paulo, and Washington DC. The first AA 1000 Framework, developed in 1999 has since become the foundation of the AA 1000 Series. The standards are intended to help organizations raise their social and ethical accountability with stakeholders' participation.[10]

ECS2000

The Ethics Compliance Management System Standard (ECS2000) is issued by the Business Ethics Research Project, Reitaku Center for Economic Studies, Reitaku University, Japan, in response to the need for an effective business ethics and compliance system. ECS2000 provides corporations with ethical standard, systematically specifying the development of management systems to ensure ethical practices.[11]

Currently, the most popular guidelines to communicate the sustainable development of CSR activities to stakeholders are the GRI reporting standards. *Sustainable reporting* is reporting on the triple-bottom-line factors; namely economics, environmental, and social policies; the impacts and performance of an organization and its products in the context of sustainable development (see figure 6.1). In an article written by Egon Zehnder, international consultants, the sustainability elements are defined very clearly:[12]

FIGURE 6.1 An ideal model of economic, social, and environmental interrelationship of sustainable development

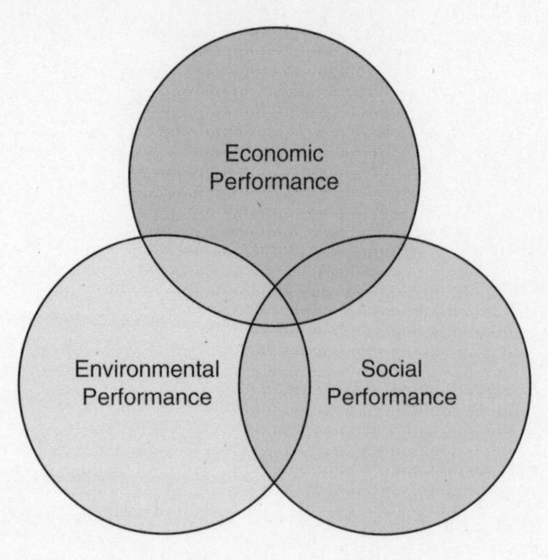

Economic sustainability is the ability of an organization to maintain economic profitability by generating value from its offering of products and services. Successful organizations find a way to translate their relative competitive advantages into economic values that deliver returns to their owners.

Environmental sustainability is the ability of an organization to ensure that's its long-run economic behavior is not undermining its own or the broader natural environment. This recognizes that natural resources have a finite capacity either for exploitation or as sinks for pollutants.

Social sustainability is the recognition that an organization operates within a broader social context and relies on it to prosper and survive. The social dimension of sustainability affects an organization's activities at every step of the value chain—from suppliers' use of international labor to employee, customer, and community engagement.

INTRODUCTION TO SUPPLY CHAIN RESPONSIBILITY

Globalization has changed how businesses manage their operations. Over the past few years, we have seen corporations engage in a cost-effective, yet complex, global manufacturing and distribution chain, involving networks of business partners in distant countries. International sourcing presents another social and economic issue that in June 2008, the Social and Economic Council of the Netherlands (the Sociaal-Economische Raad, SER) issued an advisory report on recommendations to government and industry both on the international aspects of CSR, and specifically in the area of international supply chain responsibility.[13,14]

Definition: Supply Chain responsibility is a voluntary commitment by companies to manage their relationships with suppliers in a responsible way, also referred to as responsible sourcing (SER/ICC).

The commitment is voluntary, but is not free of obligations, such as they are not legally responsible for what the international suppliers do. The effectiveness of the commitment will be tailored to the specific circumstances in each case.

With the objective of promoting good practices in the area of CSR, SER together with the peak employers associations and trade unions have agreed on the followings:

1. It was agreed that CSR should be encouraged and facilitated at all levels, including the international supply chain responsibility. The social context requires openness and transparent communication, which means that the enterprise must respond to legitimate questions and demand. Transparency enables enterprises to win the trust of the stakeholders, build their reputation, and command employees', and customers' loyalty.
2. The normative framework to identify what is expected to be done by enterprises with respect to the international aspects of CSR and supply chain responsibility are as follows:
   - The ILO Declaration on Fundamental Principles and Right at Work (1998) concerns the freedom of association, the right to collective bargaining, and the ban on forced labor, child labor and discrimination. The 1998 ILO Declaration was reaffirmed in the ILO Declaration

on Social Justice for A Fair Globalization, dated June 2008.

- The ILO's Tripartite Declaration of Principles concerning Multinational Enterprises and Social policy makes recommendations on employment, training, condition of work, life, and industrial relations (2000).
- The OECD Guidelines for Multinational Enterprises (2000) make recommendations on reporting, employment and industrial relations, the environment, anti-corruption, consumer affairs, science and technology, competition and taxes and advises enterprises to encourage business partners including suppliers and sub-contractors to adhere to the Guideline.
- The recommendations of the International Chamber of Commerce (ICC) on supply chain responsibility (2007) and the guide to responsible sourcing based on these recommendations were developed in 2008. They involve integrating responsible sourcing into the enterprise's buying practices, making expectations clear to suppliers, helping suppliers set their own business standards, tracking supplier compliance, and finally developing a policy for dealing with non-performance.

3. This normative framework should be used in the international operations, to promote good practices in the area in which the enterprises are operating. Transparency, verification, and internal complaints procedures are important, to be reported annually in a separate report.
4. SER planned to monitor and draw up an annual progress report starting in 2009 to be discussed by a special committee set up by SER, which will meet twice a year. Evaluation will be done in the second half of 2011 and evaluation will be completed by July 2012, based on the monitoring from 2009 onward.

This framework offers clear direction and guidance for businesses to pursue the creation of the three-dimensional values, which are profit, people, and planet, particularly in operating globally.

International Chamber of Commerce (ICC) Guidance for "Responsible Sourcing"

With consumers and customers increasingly basing their purchase decisions on not only product quality and safety and continuity of supply and delivery, but also on the sources of the products, that is, the ingredients and the processes used in making the products, pursuing good international supply chain responsibility is becoming a basic strategic imperative and need. For this reason, some companies have included in their integrated procurement strategy for suppliers issues on working conditions, ethical, environmental, health and safety, and human rights. Good practices in those areas could ensure significant contribution toward the ultimate goal of supply chain continuity and long-term efficiency and the certainty of maintaining reputation and or brand equity.

Currently, the competitiveness of a nation or a company to attract either direct investment or become a sourcing country and partner, as part of the international supply chain, is not only about availability and price, but also covers areas like compliance with the national law and care about community (people) and the environment (planet). In many parts of the world, the lack of government involvement in enforcing social and environment standards makes it difficult for companies to ensure that good business practices are consistent across their global supply chain.

However, companies are expected to manage their supply chain responsibly through working with their suppliers to collaborate toward improving their social and environmental performance such as: using information and training on the development of management system, carrying out assessments of suppliers' facilities and practices, and treatment of workers through on-site visits, worker interviews, and independent monitoring where appropriate.

It is recommended by ICC in measuring suppliers' social and environmental performance to use the Global Compact and its "10 principles" covering human rights, labor standards, the environment, and anticorruption (www.unglobalcompact.org), which is increasingly being recognized by companies as a benchmark for good business practices.

THE MOVEMENT TOWARD ISO STANDARDIZATION OF CSR

Considering the many CSR standards available in the world, during a meeting in September 2002, the ISO council decided to request that the ISO Technical Management Board (TMB) create a high-level advisory group on CSR for the purpose to examine whether the ISO should develop the CSR Standards.

ISO, or the International Organization for Standardization, is a worldwide federation of national standards bodies (ISO member bodies). The work of preparing the international standards is normally carried out through ISO Technical Committee. International Standards are drafted in accordance with the rules given in the ISO and IEC Directive. Draft No. 4 of the International Standard for CSR (ISO and CD 26000), adopted by the technical committee, was circulated to the member bodies for voting.[15]

With the issuance of committee draft (CD), expert's participation in each member country will be channeled through the National Standard Body who manages stakeholders meeting and mirror committee of Social Responsibility including Indonesia. *The International Standard for CSR (ISO and CD 26000) is expected to be finalized by 2010.*

There are seven principles of social responsibility that the ISO 26000 document promotes, namely:

- accountability
- transparency
- ethical behavior
- respect for stakeholders' interests
- respect for the rule of law
- respect for international norms of behavior
- respect for human rights.

These principles are to be applied on what the document refers as the seven core subjects, with the organization and its governance as the key element (see figure 6.2). When adopting the standard, the organization should take into account the integration and interdependency of these core issues.

FIGURE 6.2 The seven core subjects promoted by the ISO 26000 document

Redefined in the document, community development is now engaging community participation, and is named "Community Involvement and Development," basically highlighting community involvement, education and culture, employment creation and skills, technology development, wealth and income creation, and health and social investment. It is the responsibility of an organization to understand and prioritize the problems and needs of the community, get the maximum participation with the aim to empower them.

PROGRESS OF LEGISLATION AND REPORTING IN INDONESIA

Several initiatives marked the country's awareness and growing commitment to sustainable development.

Within the past three or four years, the Indonesian government has also been proactive in pushing forward the CSR agenda. The Indonesian Limited Liability Companies Law (Law No. 40 and 2007), which was enacted in August 2007 includes a chapter that stipulates environmental and social responsibility (ESR) as an obligation for corporations operating in the country. Implementation of ESR is mandatory for businesses in the fields related to natural resources. With this new law, *Indonesia is becoming the first country (2007)* to legislate CSR as mandatory obligation.

However, the required detailed guiding rules of CSR definition and its implementation framework still need to be issued. Any law and regulation requires detailed guidance of definition and implementation. Without which, implementation and enforcement could not be done. Therefore, until now implementation of CSR is still done voluntarily. Nevertheless, this new law obliges companies to allocate and calculate CSR as a cost of the company.

The Association of Chartered Certified Accountants (ACCA), in collaboration with the Ministry of Environment, has issued a guideline on sustainability for the Republic of Indonesia, that is, *An Introduction to Sustainability Reporting for Organizations in Indonesia.* On August 9, 2004, the Indonesian Institute of Accountants Management Accountants Compartment (IAI-KAM) organized an international conference, "Sustainable Enterprise Performance Conference (SEPC 2006)." The theme of the conference was "Creating Value and Meeting Stakeholders' Expectations."

The result of this conference was the the establishment of National Center for Sustainability Reporting (NCSR) in 2005. Five prominent independent organizations, the Indonesian Institute of Management Accountants (IAMI), the Indonesia-Netherlands Association (INA), the National Committee on Governance (NCG), the Forum for Corporate Governance in Indonesia (FCGI), and the Association of Indonesian Publicly Listed Companies (AEI), pooled their resources in this initiative with a vision to become the leading organization in providing sustainability reporting standards for corporations in Indonesia.

To support its objectives in promoting the sustainability management and to elevate sustainability reporting practices in the country, NCSR has also structured working programs covering the areas of interorganizational relations, training and education, research and consultancy, and development and dissemination of reporting guidelines to companies and organizations in Indonesia. The purpose of these programs is to have all Indonesian publicly listed companies report their CSR activities by 2010, and to make a stand-alone sustainability or CSR report is mandatory by 2013 for organizations with high social and environmental impact.

The newest development is the launch of the First South East Asian Country SRI Index (June 2009) to measure the performance of Indonesian companies adopting sustainable business practices by the Indonesian Biodiversity Foundation KEHATI, supported by IDX, the Indonesian Stock Exchange. This is useful as an additional benchmark for investors to track and evaluate their investment especially in well ESG-managed (environmental–social–governance) businesses. To reduce the performance risk, now responsible investors have a tool to select companies for their investment, that is, the integration of best practices in ESG into the selection portfolios.

Indonesian companies also get a clear incentive to be included in the KEHATI-SRI Index to attract:

- investors who want to invest into sustainable businesses
- the best employees who want to work for the best companies
- clients who want to mitigate risks in the supply chain

KEHATI is a nonprofit, independent organization whose key role is to facilitate support and to motivate various parties; the government, community, and private sector in biodiversity conservation and biological management. KEHATI is the owner of SRI-KEHATI Index, and with the help of OWW Consulting—a CSR solution and SRI research firm based in Singapore and Kuala Lumpur—decides the rating criteria and selects the companies to be included in the index. Review is done twice a year. The Indonesian Stock Exchange provides a new perspective of the country's commitment to responsible investment. This special index is calculated and maintained by the Indonesian Stock Exchange, while OWW Consulting provides the underlying ESG methodology and ratings of the companies.[16]

NOTES

1. http://www.bized.co.uk/learn/accounting/management/social/index.htm.
2. New Economics Foundation.

3. This segment is taken from the original document published by the Council for Better Corporate Citizenship (May 21, 2002; revised January 14, 2003), *International Corporate Social Responsibility (CSR) Standards and Norms: Present Situation, Future Challenges.*
4. See http://www.globalreporting.org/AboutGRI/.
5. See http://eurlex.europa.eu/LexUriServ/site/en/com/2001/com2001_0366en 01.pdf.
6. See http://www.cauxroundtable.org/index.cfm?&menuid=2.
7. See http://www.unglobalcompact.org/AbouttheGC/.
8. See http://www.oecd.org/pages/0,3417,en_36734052_36734103_1_1_1_1_1,00.html.
9. See http://www.saintl.org/index.cfm?fuseaction=Page.viewPage&pageId=473.
10. See http://www.accountability.org.uk.
11. See http://www.ie.reitaku-u.ac.jp/~davis/html/ecs-eng-main.html.
12. Christoph Lueneburger and Richard Murray-Bruce, *Sustainability Leadership: Making Change Happen* (Egon Zehnder International: 2009), (http://www.egonzehnder.com/global/practices/functionalpractices/sustainability/thought leadership/article/id/83700053).
13. *Statement on International Corporate Social Responsibility*, Sociaal-Economische Raad, 2008
14. Ibid.
15. Indonesia is also a member.
16. http://www.sri-asia.com/products/kehati-sri-index-2.html.

PART 2

CASE STUDIES

The following section profiles nine corporations of various industries applying corporate social responsibility (CSR) initiatives of various perspectives. Specific programs and innovations carried out by these companies are described in detail to show how these have offered long-term benefits to the community and competitive advantage for the company. In some, the impact has served a much wider interest, including that of a nation and the world. *The contributions of these companies are well recognized by all the awards they have received.*

The cases presented describe the evolution of CSR within the different stages of community and country development—emerging, developing, and developed markets, covering multinationals: Unilever, Heinz, TNT, Intel, Motorola, and ITM, as well as local Indonesian companies: PURA Group, Bank Danamon, and Astra International. Stories of two of the multinationals, Unilever and ITM, illustrate practices applied in the Indonesian market and environment, whereas the rest are CSR implementations in different countries. Each of the cases demonstrates how CSR has become part of the company's operations and has proven to give the company and the society sustainable benefits. Readers will also appreciate a unique case of sustainable agricultural practice with high social and economic impact implemented through projects initiated by the royal family of Thailand.

To understand the cases better, where it starts with Unilever in the setting of Indonesia as an emerging market in the 1970s, we will agree that to ensure a successful entrance into an emerging market, the following activities have to be done: introduction of socially innovative products (identifying customers needs at the right timing) combined with efforts to create, condition, and expand the market through community development and education to improve lifestyles. At the same time, the establishment of a value chain, from supply of raw and packing materials to distribution and sales, have to be developed from scratch.

It has to be understood if you enter an emerging market, wealth and job creation to ensure a conducive environment for growth and sustainability is very important.

Earlier in part I we have seen the idea applied by Unilever Indonesia, Aqua, and PT Sinar Sosro (*Teh Botol*),[1] which had recognized a gap in the market and introduced socially innovative products, supporting the introduction with community education for market development and the establishment of the value chain. Without realizing it, these companies had been implementing the fundamentals of CSR. Up till now, they still enjoy their market leadership and each of the brands is becoming the generic name of the product category.

Identifying the right CSR strategy, as Unilever, Aqua, and PT Sinar Sosro had previously done, is critical to companies operating in emerging markets to gain long-term competitive advantage. This concept, however, is also applicable to developing and developed markets, especially with the advancement of communication technology and traveling that has led to the world becoming one global market.

In fact, companies in developing countries are becoming part of this wider market, and globalization has opened further opportunities for multinationals to take advantage of capabilities located in other parts of the world. Furthermore, because of cost increases of raw materials and wages in the home countries coupled with the competitive development of a wider global market, many companies from the developed countries seek greater efficiencies in their supply chain by establishing global or regional sourcing facilities, innovation centers or other centers of excellence in developing countries. All these offer developing countries prospects of increased revenue, employment, and knowledge, but also pose multidimensional challenges for businesses: from human resources development, training, capability; operational excellence and governance; customers' needs and satisfaction to the changing government regulations, and the emergence of nongovernmental organizations (NGOs; see chapter 1, figure 1.1).

A country's civil foundation tends to grow in line with its overall economic development. The different stages of the world's civil foundations can affect the global supply of corporate responsibility positively or negatively. On the positive side, global corporations from advanced countries can enter developing economies and bring with them the employment, ethical and environmental practices of their home countries' civil foundation. In adopting those practices, local businesses engage in responsible behavior that eventually could add to the country's civil foundation. In this situation, globalization can "average up" the world civil foundation. On the other hand, when a corporation from an advanced country does business in a developing country, it may establish a level of corporate virtue, in line with the host country's civil foundation. This is considered as "averaging down" its level of corporate responsibility.[2]

Within a current capability-based competition, companies do not operate in isolation from the society around them. In fact their ability to compete depends very much on the environment, whether or not it is conducive for their operation. The more social improvements relate to a business, the more it leads to economic benefits as well. This is true for all companies operating in both developed and developing countries.

With the advancement of technology the "capability-based" competition goes even further in terms of "modern knowledge and technology-based" competition, which depends more on workers' capability. So a company's main assets have evolved from its building and machinery into its people. To reduce innovation time in introducing new products and preempt competition in a certain country, companies have established local innovation centers in cooperation with local universities or research institutes and local suppliers. They collaboratively conduct research and development in close cooperation with customers to better understand their needs. Managing the increasingly complex local regulations and reducing approval times for new projects and products are also becoming part of a competitive edge of a company. Globalization has made international companies depend more on local partnerships, and they rely on outsourcing and collaboration with local suppliers and institutions rather than on vertical integration.

There is one other major issue concerning capability-based strategy in the context of global structure. It is very important to mitigate possible risks to company reputation or brand equity, particularly for companies with brand exposure and consumer visibility. To preempt possible reputation risk, the basic principles of the three Ps should be adhered to. Not only is cost effectiveness (profit) important, but care about community (people) and the environment (planet) is also a must, contributing to averaging up the social responsibility within the developing country in which the company is operating.

Because the current competition depends very much on people, capability, and the fact that people and workers have evolved to become a company's main assets, it is clear that successful CSR activities should focus on people within and outside the companies. Specifically, CSR programs should work in the areas of people welfare, safety, education, community capacity building, employment, and improvement of community's quality of life, or wealth creation, which will give sustainable benefit to companies and the people/community.

A good example is the familiar case of Nike, a company known for its commitment to environmental caring initiatives, but failure to interpret the fundamentals values of all the three Ps has cost Nike's reputation hugely. Bear in mind that at the time when the case emerged, no global standard measures on sustainability or social responsibility had been defined. One is never able to predict the boundary of a reputation risk, both the timing as well as the cost (see the Nike Case below).

The Nike Case: The Labor Issue

An example of problems associated with "Company Reputation Risk"[3]

It started from Nike's originating business model, specifically to produce in the most cost-effective way (in unregulated developing countries) high-quality expensive shoes to be marketed in highly developed markets.

Nike had developed and adopted a comprehensive environmental policy before one was forced to do it, which was supported by a full-fledged department—Nike Environmental Action Team (NEAT) in 1993—focusing on "green" issues of recycling and environmental education.

Nike has always been very committed to producing high-quality products and consequently quality has always been strictly controlled and stipulated in details in its contract with suppliers. However, there was never much attention paid to the working conditions at the production sites of its suppliers. These two policies are considered very contradictory, especially when people believe that environment is not only associated with green issues, but also covers community and people welfare.

This labor issue first surfaced in 1988, when an Indonesian trade union newspaper published a detailed investigative report exposing

(*continued*)

(*continued*)
poor working conditions at a South Korean-based shoe company producing "running shoes" for Nike. Using cheap overseas labor to create products at low cost does not necessarily mean competing globally. Although Nike obtained high-quality products and good profit, it was balanced out by the negative publicity, because many responsible consumers do not feel good about running in footwear that might have been produced by workers whose rights were disregarded.

A few other negative publications have also appeared in other Indonesian newspapers, reporting on wage protests and bitter strikes at several local Nike subcontractors, after which the Indonesian daily *Media Indonesia* ran a three-day report on abuses at Nike shoe factories headlined "World Shoe Giant Rapes Workers' Rights."

By 1991, Western media such as Thames TV in the UK, *The Economist*, and the *Knight Ridder* chain of newspapers all filed reports on poor working conditions at Nike contractors in Indonesia. The peak of all this negative publicity was a lengthy report on Nike's Indonesia operation appearing in the company's local newspaper, the *Oregonian*. This was proof that bad publicity was traveling very fast all over the world, even before the internet era. Any wrongdoing in one part of the world will affect company's reputation worldwide.

After this series of negative publications, Nike undertook many corrective actions. It formulated its first "Code of Conduct and Memorandum of Understanding" for contractors, and hired accounting firm, Ernst & Young, to do its first "social audits" of the conditions at its Indonesian contract factories. In 1997, Nike hired a former UN Ambassador, Andrew Young, to tour the Asian factories in the hope that a clean bill of health from a well-known civil rights advocate would improve Nike's public image at home and abroad.

Furthermore, Nike has done a lot in developing safe raw materials in their products. Nike aimed to become the world's first shoe company to eliminate 100 percent of the polyvinyl chloride (PVC) used in its sneakers, has also analyzed the toxicity of every chemical used in all plants all over the world to ensure elimination of all dangerous chemicals from its manufacturing process, and has worked on many other responsible actions. Still, all these efforts didn't stop the negative publicity. Although many other international companies were also moving their production overseas, subcontracting it out, Nike was still condemned as giving a bad example of globalization.

Even after years of adopting a culture of responsibility, involving serious commitments to improve basic practices and processes since the problem first surfaced in 1988, Nike still is struggling to find the right solution to *reinstate* fully its brand image and company reputation.

This case study is an example where competitive edge could only be achieved if you simultaneously aim to target at all three Ps (profit, people, and planet), while in the case of Nike their first business model was only aiming at two Ps (profit and planet). It also illustrates that it is better, and surely less costly, to preempt or mitigate reputation risk, than to try to reinstate a tarnished reputation or brand equity. Within the knowledge economy, sustainability must extend to people as well as to the environment.

Both CSR and good corporate governance are becoming very important basic business strategies, especially when customers or consumers become more critical and are better educated—through the global communication and the internet—and also becoming more aware of their rights and their potential power to influence corporate behavior. *Increasingly, consumers' purchasing decision considers the sources of the products they buy, including the ingredients and processes used in making the products, therefore being a good corporate citizen is fast becoming a competitive edge.*

The CSR practices consider the three dimensions of value creation: economic (profit), social (positive consequences for people inside and outside the company), and environmental issues (planet), simultaneously targeting areas of competitive context, which ensure sustainable benefit for both the company and society. Although it cannot replace the government in providing public services or infrastructure, CSR activities, especially in a developing country, could have the potential to contribute to the acceleration of microeconomic growth through companies using good governance, human resources training, operational excellence, and best practices. There are many companies that participate in public–private partnership initiatives around the world, dealing with social issues such as capacity building and education, health, safety, and local economic development.

Specifically within a developing country in which failure to implement good corporate governance more widely could result in costly operations and will reduce its competitiveness to become a sourcing country, particularly with the liberalization of new industrial countries such as China. Currently, more and more local companies recognize these new challenges, and they also realize that ensuring the competitive edge of the nation is a joint responsibility of the government and the local and international corporations operating in Indonesia.

Although good corporate governance is still sporadically implemented, some prominent Indonesian businesses and institutions realize the importance of good governance to survive within the global competitive market, particularly after the crisis years of 1997–98.

So most of them are already operating as islands of integrity and adopting governance principles within their operation. Also, CSR principles integrated within their internal behavior and culture is becoming a critical part of those company's core competence and strategy in spreading good governance and best practices to all stakeholders and business partners. As already mentioned in part I, this approach has been effective in mitigating the effect of the 2009 global crisis, because Indonesia's GDP is still growing at 4.5 percent, compared to the negative growth of the neighboring countries.

Discussed later in the case studies is what selected companies have already done to contribute to the education of people and the community in Indonesia; to create employment or self-employment during scarce job availability outside their value chains. Although each of them has implemented very successful CSR activities, collaborative efforts, cooperation, and communication among the companies and also exposure in media are relatively still very limited.

Ideally, corporations' collaboration, particularly on leveraging technical capabilities supported by the use of available corporation's facilities for workforce training—if managed effectively—could give substantial contribution to the national development of skilled/trained labor force, which is one of the most important assets in attracting foreign direct investment. Companies developing together schools for vocational studies and giving workers on the job training in the factories or premises of companies will have a twofold benefit. Better skilled labor will improve effectiveness of companies and during scarce job availability these people are still able to earn through self-employment. The problem of outsourcing—which has become "a labor issue in Indonesia"—could also be solved by giving labor the same chance to upgrade their skills through vocational educational training. With higher-efficiency output, wages at outsourced companies (small and medium-sized enterprises) could also be increased accordingly.

Furthermore, competitiveness today depends on the efficient productivity with which a company uses its labor, capital, and natural resources in producing high quality products and services. Implementing both good corporate governance and CSR within their operations, that is applying the three-dimensional approach (profit, people, and planet) in creating new products, is equal to spreading technology and raising productivity, increasing quality, and improving services. By doing this, businesses could become active agents in employing and socializing operational excellence and business ethics within their nation.

A very good model of effective use of joint CSR budget for education is what Sampoerna Foundation in Indonesia has done for tertiary education. The foundation established the necessary infrastructure and services, including identification of educational needs in Indonesia, to be shared with partners and other beneficiaries.[4] This system can also be done for vocational studies, covering the majority of the population, to create employment and wealth, and consequently improve the quality of life of the community.

To give a bird's eye perspective of high-impact CSR approaches, which are completely integrated into the core business practices, giving sustainable results, the following gives a quick review of each of the case studies, particularly focusing on community capacity building, empowerment, and job creation with the exception of CSR in Thailand.

- *CSR in Thailand.* A study published in the *Journal of Corporate Citizenship* (2006)[5] revealed that CSR is still at an initial stage in Thailand. The finding indicates no complete integration of CSR in the core business practices of most leading companies. None has specific policies on CSR. Even so, CSR is a concern in almost all companies. Although the interpretation of CSR (definition, methodology, implementations) varies, the principles of CSR are already rooted in Thai corporate culture and practice.

Unlike its neighbors in Southeast Asia, Thailand has never been colonized, so the development of sustainable community capacity building and empowerment and creation of employment and wealth are going through a different route, such as the royal family projects, which provide good examples of social responsibility for Thai and international corporations, particularly in focusing on sustainable development: providing higher incomes for the community while protecting environment and conserving important social values.[6]

- *Unilever India (Hindustan Lever) and Unilever Indonesia* have already implemented the philosophy of CSR as an integral part of the company's strategic framework: from the 1950s and 1960s for India, mid-1970s for Indonesia, even before the trendy acronym of CSR was known.

How resilient a company is, with good reputation and well-accepted goodwill of its brands, was tested during the time of the financial and political crisis in 1997–98. During that particular period, Unilever Indonesia had adapted its business model by ensuring consistent high quality of all products, yet still remaining affordable (socially caring) through:

  - discarding unnecessary packaging, which was selling toothpaste in tubes without cartons; 6 tubes were plastic shrink-wrapped for protection during transportation
  - expanding distribution in traditional outlets from 240,000 to 350,000 (easily reachable by all population including semi urban and rural areas)
  - strengthening further partnerships with all parts of the value chain to ensure sustainability
  - prioritizing the retention of highly motivated employees (three times of salary increase with total increase of 55 percent versus inflation of 80 percent)
  - expanding the local operations through joint ventures and acquisitions

Growth of sales and income of the year 1998 were 70 percent and 20 percent, 10 of the 11 product categories had increased market shares, with no disruption in production and no laying off of people. It took only five years for Unilever's sales—in US dollars—to return to the 1996 level (1996: $1 = IDR2000 versus 2002: $1 = IDR 9000).

The resilience of the good Unilever reputation and the good equity of its brands were proven during the special circumstances of the 1997–98 financial crisis, by comparing Unilever Indonesia's performance relative to other big companies (FMCG) between 1998 and 2002. Sales in Indonesia's FMCG market fell by nearly 22 percent. while Unilever sales for the same period more than doubled, growing by 137 percent from $319 million to $757 million.[7]

- *Astra International* has, through its foundation, created programs on mechanics training and education and other income-generating activities, including entrepreneurship training for retirees. These people will be able to become full-fledged trained mechanics, opening doors of opportunity to becoming self-employed service mechanics, working in association with spare parts retail shops, or they are able to open their own service stations for motor car or motor cycles. By systematically providing practical skills to productive workforce, Astra has helped create new wealth at the same time these trained independent mechanics will ensure certainty of very good after-sales service to customers and users of Astra products. Most of them are becoming loyal users of Astra's motorcycles. These activities have ensured maintenance of Astra high reputation and market leadership.
- *PURA Group* has started, in 1975, an education program for the local community to socialize the benefit of paper waste collection and becoming a supplier for PURA. Paper waste collection is becoming a business for many small companies, and they have developed into reliable suppliers of many local paper manufacturers, even exporting it to paper industries outside the country. PURA Group has changed the very low value of husk—waste products of rice milling plant—to become a useful fuel material with high economic value, through modifying the burner of its electric power station boiler to run not only on coal, but also on coal and husk. Husk is supplied by local community, so it is creating sustainable employment and wealth

(PURA needs 5,000 tons of husk per month and that is expected to rise along with the growth of the company).

- *Bank Danamon* is a one of the leading commercial banks in Indonesia, with a high commitment to supporting micro- and small-scale entrepreneurs. Its innovative products, designed to cater this specific segment, the *Danamon Simpan Pinjam* or self-employed mass market, are well accepted by microbusinesses that have no access to a banking system. *Danamon Simpan Pinjam* has grown to more than 850 branches located in more that 1,500 traditional markets, where an estimated 12.6 million microentrepreneurs rely on the program for their livelihood.

  These traditional markets have been facing severe competition from the modern trade—supermarkets and minimarkets—penetrating their neighborhoods. To ensure the survival of the traditional markets and strengthen its competitive edge, Bank Danamon, through its foundation (Danamon Peduli), has implemented an integrated CSR program to improve the quality, hygiene, and cleanliness of the traditional markets. Under this program, called My Clean, Healthy, and Prosperous Market, the foundation collaborates with local governments and a biotechnology institute. Together they created simple waste management and processing systems for the traditional markets. Clean markets and the huge amount of market waste, which are processed into high-quality organic fertilizer, provide economic benefits to traditional vendors.[8]

  Furthermore, capacity building and training in simple financial management for the self-employed and small-scale vendors are carried out to give a competitive edge to the microentrepreneurs. These activities have secured the reputation of the company and a positive impact on the growth of its business.

- *ITM* has complied with all government regulations and, at the same time, has done community development and education, creating sustainable employment and wealth—even after its mining operations are closed—within the surrounding areas of its mining operations. As a result, it has gained wide acceptance and respect from the central and local government, the community, and NGOs. Its financial results are growing as well.

All these successes of having mutually beneficial initiatives are no different from those achieved by companies implementing CSR programs outside Indonesia.

- Because of its proven long-term commitment to the welfare of the society, it is worth mentioning *Tata,* even though no specific case study is dedicated to its discussion in the following pages. Tata is a leading conglomerate in India, with more than 100 companies operating in more than 80 countries and every major market in the world. It has a 140-year tradition of community and national development. Social responsibility is a way of life for the Tata Group of Companies, a reason many Indians are proud of Tata's achievements and services. By the late-nineteenth century, Tata had started to grant scholarships to Indian students for higher education studies abroad. Tata's commitment to education, the community, ethical business, and the environment is supported by setting up the Tata charitable trusts, which now control 65 percent of the holding company's shares. and the many sustainability initiatives carried out by the Tata companies.

  Tata (Tata Motors) has also recognized the importance of women and youth's role in accelerating the development of society. Through vocational training and capacity building to increase their skills, the company is improving the quality of life of women and youth in India's rural areas, because with their acquired new skills, they are able to be independent and earn their own money. To ensure the success of this project, the company is providing special training facilities, for example, beautician courses and tailoring for women and training for motor mechanics, electrical, AC repair, and refrigeration for youths.

  The company has also empowered the community with entrepreneurial skills through cooperatives and has initiated and guided the formation of several cooperatives, such as the Chaitnya Cooperative Society for recycling wood, which employs 350 members, and the Sahajeevan Cooperative Society, which employs 250 people. The success story to note is the Tata Motors Grihini Welfare Society (TGU), established in 1974, gives employees' family members, particularly women, the opportunity to develop their entrepreneurial skills and independently earn their own income. The TGU is the only women's cooperative society to receive the ISO 9001–2000 certificate from Bureau Veritas Quality International for its cable harness and electronic units.

To ensure successful implementation of its CSR activities, Tata is continuously doing capacity building and community empowerment through training and providing the necessary infrastructure and seed capital. The company also treats these societies as favored vendors by making them their suppliers.

- At *Heinz*, its vision drives the company's efforts toward a broader meaning, the sustainable health of people, the planet, and the company. Heinz ensures its competitive advantage by applying one of its key sustainability programs: the Heinz-Seed program. Under this program, Heinz supplies premier hybrid tomato seeds annually to farmers around the world, for them to grow higher-yield, high-quality tomatoes with reduced fertilizer. Heinz-Seed is now a model for sustainable agriculture globally, and is a market leader. Heinz and its foundation have also launched the Heinz Micronutrient Campaign to combat iron deficiency and vitamin and mineral malnutrition in the developing world. Heinz has been working closely with local governments, NGOs, and other international and domestic health organizations to reach 10 million children at risk of micronutrient malnutrition by 2010.
- *TNT*, as a leading global transport and logistics company that operates one of the world's largest air and road fleets, is also in the lead in the reduction of CO_2 emissions efforts. Under a program called Planet Me, the company aims to ensure that TNT manages its contribution to the effect of global warming effectively. TNT has also been an active partner of the United Nations World Food Program, helping organizations to fight world hunger. This program has given the company undisputable reputation.
- *Intel*, a leader in semiconductor technology has gained its competitive edge through innovation and excellence. This is the reason Intel is very serious in spreading knowledge by investing more than one billion dollars in community programs and resources to help teachers, students, and universities around the world in the areas of math, science, and technology.

- *Motorola* is a pioneer of mobile communication, and has been in the business for almost 80 years. Motorola's commitment to corporate responsibility is well recognized and the company was ranked as Sustainability Leader by Dow Jones Sustainability Index in 2008 for the fifth time. In managing its complex global supply chain, Motorola has also invested in and developed suppliers in developing countries, generating new jobs and wealth opportunities. Motorola, which has been bringing communications to people in isolated places, has also invested in community development programs and gaining acceptance that further boost the company's reputation.

It is proven that a successful integration of financial and social objectives, gives businesses a competitive edge.

> "Contrary to business school doctrine, "maximizing shareholder wealth" or "profit maximization" has not been the dominant driving force or primary objective for visionary companies. Visionary companies pursue a cluster of objectives, of which making money is only one—and not necessarily the primary goal. Yes, they seeks profits, but they are guided by core ideology, values, and a sense of purpose beyond just making money, Yet paradoxically, the visionary companies make more money than the more purely profit-driving comparison companies.[9]"

NOTES

1. See chapter 3, "Fundamentals and Evolution of CSR".
2. Roger L. Martin, "The Virtue Matrix: Calculating the Return on Corporate Responsibility", *Harvard Business Review,* March 2008, Vol. 80.
3. Jeffrey Hollender and Stephen Fenichell, *What Matters Most: Business, Social Responsibility and the End of the Era of Greed,* Perseus Books, December 2003, pp. 184–201.
4. See chapter 4, "How to Optimize CSR".
5. Suthisak Kraisornsuthasinee and Frederic William Sarerczek, "Interpretations of CSR in Thai Companies", *Journal of Corporate Citizenship,* 2006, http://www.britannica.com/bps/additional/content/18/22097290/Interpretations-of-CSR-in-Thai-Companies.
6. See case study 10, "CSR in Thailand".

7. Jason Clay, *Exploring the Links between International Business and Poverty Reduction: A Case Study of Unilever in Indonesia* (An Oxfam GB, Novib, Oxfam and Unilever Indonesia joint research project, Oxfam GB, Novib Oxfam, Netherlands, and Unilever, 2005).
8. This Danamon Peduli CSR program "Nothing Wasted—Indonesia" was selected as the second winner of the BBC World Challenge 2009 award—The Hague, December 1, 2009.
9. Jim Collins and Jerry Porras, *Built to Last* (HarperCollins Publishers, 2004).

Case Study 1

PT UNILEVER INDONESIA, Tbk

THE COMPANY

PT Unilever Indonesia, Tbk, or Unilever Indonesia, a subsidiary of the Unilever Group,[1] was established in Indonesia in 1933. Unilever Indonesia has grown to become one of the leading suppliers of fast-moving consumer products across foods and ice-cream, and home and personal care categories. Its portfolio includes many of the world's well-known and well-loved brands such as *Pepsodent*, *Pond's*, *Lux*, *Lifebuoy*, *Dove*, *Citra*, *Sunsilk*, *Clear*, *Rexona*, *Vaseline*, *Rinso*, *Surf*, *Molto*, *Sunlight*, *Walls*, *Blue Band*, *Royco*, *Bango*, and many more.

Unilever Indonesia has two subsidiaries, PT Anugrah Lever and PT Technopia Lever. PT Anugrah Lever[2] is engaged in manufacturing, developing, marketing, and selling sweet soy sauce, chili sauce, and other sauces under the *Bango* trademark and other brands under license. PT Technopia Lever, with 51 percent of its shares owned by Unilever, is distributing, importing, and exporting goods under the *Domestos Nomos* trademark.

In 1981 and 1982, the company was listed on the Indonesian Stock Exchange and, by 2007, is ranked tenth in terms of market capitalization. Having its shares registered and traded publicly, Unilever is strongly committed to continue advancing together with Indonesia. In 2007, Unilever Indonesia achieved a 14-percent net profit growth or IDR2 trillion, with 11-percent sales growth or IDR12.5 trillion. Supported by eight factories in Jababeka (West Java) and Rungkut

(East Java), Unilever sells its products through a network of about 400 independent distributors to reach hundreds of thousands of outlets throughout Indonesia. Products are distributed through distribution centers, satellites warehouses, depots, and other facilities.[3]

CORPORATE GOVERNANCE AND CORPORATE SOCIAL RESPONSIBILITY

Unilever has earned a reputation for conducting its operations with integrity and with respect for people, organizations, and the environment. The company seeks to manage and grow the business in a responsible and sustainable fashion. The values and standards by which the company expects to be judged are set out in its *Codes of Business Principles,* shared with its business partners including suppliers and distributors.

The following is the Unilever Global Corporate Purpose, which supports the company approach to governance and corporate responsibility:

1. Our purpose in Unilever is to meet the every need of people everywhere—to anticipate the aspirations of customers and consumers and to respond creatively and competitively with branded products and services which raise the quality of life.
2. Our deep roots in local cultures and markets around the world are an unparalleled inheritance and the foundation for future growth. We will bring our wealth of knowledge and international expertise to the service of local consumers—a truly local multinational.
3. Our long-term success requires the highest standard of corporate behavior toward employees, consumers, and the societies and the world in which we live.
4. This is Unilever's road to sustainable, profitable growth for business and long-term value creation for our shareholders and employees.

The company's commitment to internal corporate governance under the third purpose is the foundation for all Unilever's externally focused corporate social responsibility (CSR) activities. The first purpose is the key to Unilever's programs on community development and its long-term result of expanding and conditioning the market, while good understanding of local tradition enables the company to form a win–win cooperation through partnerships with many small and medium-sized enterprises (SMEs), sharing knowledge and business expertise. Unilever Indonesia has defined its corporate purpose to mark the company's obligation to contribute positively to the society, and having this philosophy as an integral part of the business strategy since the 1970s has helped Unilever Indonesia maintain its significant growth for more than three decades, creating long-term value for all stakeholders and its employees.

Unilever Indonesia had implemented the fundamentals of CSR from early 1970s, which basically were market and community development:

- developing the market by creating consumer demand through market and community developments, that is, activities involving community building and improving lifestyle which in turn increases users and expand the market
- developing the product through innovation and marketing its products within an expanded market, while at the same time establishing the necessary infrastructures, that is, sourcing facilities, suppliers, logistics, and distribution network through partnership and co-operation with SMEs (social impact of the corporation through the development of the value chain)
- by involving business partners to sign *codes of Business principles*, Unilever Indonesia was promoting a range of ethical, social, and environmental standards (using best practice and good corporate governance)

All these activities were embedded within its business strategy and were becoming a strong base for growth of the company within Indonesia's emerging or developing market.

Currently, the activities under market and community development above are considered to be CSR. In 2000, Unilever Care Foundation (*Yayasan Unilever Peduli*) was established to strengthen Unilever's commitment to sustainable development, with the same end result of maintaining competitive edge, while continuously ensuring the businesses' commitment in community building, creating employment and wealth, as well as caring for the environment.

Under the *Unilever Peduli Foundation*, with its new vision, all community programs are formalized and divided into four main categories:

- CSR is an integral part of the company's business.
- Unilever considers CSR as the impact of the company business operations on the community, that is, impact through the value chain, impact from operations, and voluntary contribution.
- CSR is embedded into the way the company runs its business.
- Business and community live side by side. Doing business and caring for the community are not separate actions.

The four legs of Unilever's CSR programs in figure CS1.1 show the distribution of activities by development segments. Public health and education activities, which basically is equal to market development and expansion, and SME development programs, which started in the 1970s, are continued by the foundation with new programs catering to different lines of products and consumers.

Unilever Indonesia understands that it is the responsibility of all stakeholders, including corporations, to encourage a safe market

FIGURE CS1.1 The four legs of Unilever's CSR programs

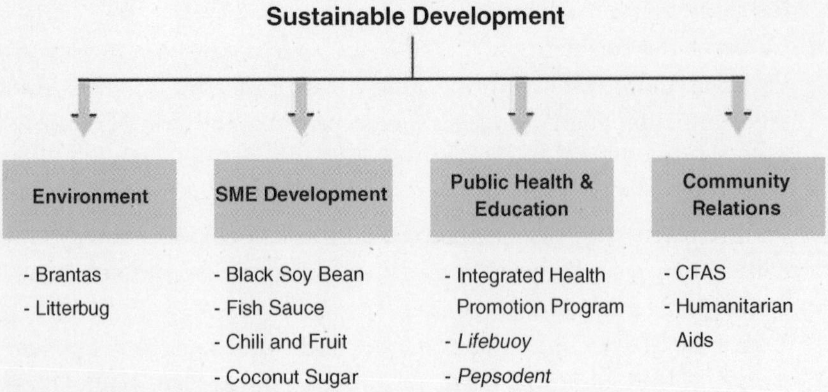

Note: CFAS=Care For Area Surrounding (the Unilever's community activities around Unilever's premises)

Source: PT Unilever Indonesia, Tbk

conducive for business to grow. With SME development programs and market expansion (through public health and education) as the driving factors behind the business' success, Unilever Indonesia can move on with the environmental projects to various philanthropic activities to give something back to society. All these activities show the company's commitment as a strong supporter of society, which ensures the corporation's positive acceptance and trust, and ensures a strong foundation for sustainable growth.

CORPORATE AND JOINT CSR ACTIVITIES DONE BY UNILEVER

Launch of Oxfam—Unilever Joint Research Report

The report, titled "Exploring the Links between International Business and Poverty Reduction, A Case Study of Unilever in Indonesia—Principal Author: Jason Clay," discussed the following:

Four areas of research:

- macroeconomic impact
- employment impacts
- the value chain from supply to distribution
- the implications of Unilever Indonesia's operation in the marketplace

The result:

- 300,000 jobs created
- two-thirds of the value generated were distributed to the participants other than Unilever Indonesia (manufacturers, suppliers, distributors, and retailers)
- taxes paid accounting for 26 percent of the value generated in the chain

The challenges:

- wealth creation not being the same as poverty reduction
- further extension to involve social institutions and resources, such as credit and saving schemes

- the closer involvement of the workers in the value chain with Unilever Indonesia's operations, thereby empowering the workers
- workers getting more benefit from the company
- setting a higher standard and how to proceed further to become even greater in the future

Asian Forum on CSR (September 8–9—Jakarta, Indonesia)

- 418 participants from 18 countries, representing 247 organizations
- 60 speakers and moderators from 15 countries
- Impressive Unilever Indonesia CSR presentation with excellent feedback from participants

Green Cities Green Communities (GCGC)

- Organic waste collection program
- Partnership based on environmental program, involving:
  MNC (Unilever's *Lifebuoy* soap and Unilever Peduli Foundation):
  - customer (*Carrefour*)
  - media (JDFI = partner of Unilever)
  - nongovernmental organization (NGO; *Kirai*)
  - government (the Ministry of Environmental Care and the Public Cleaning Services)
- Results:
  - average 4 tons of waste collected every week
  - waste collected converted into usable goods by NGO
  - three talk shows with high exposure on *Lifebuoy* and the environment
  - improved image for Unilever and the brand

Community Connection

Unilever's employees spend half a day with surrounding communities, with the following results.

- Seven Community Connection activities were conducted in 2005 alone
- Eleven schools were renovated
- Nine hundred employees were involved

A. DEVELOPING AND EXPANDING THE MARKET THROUGH COMMUNITY DEVELOPMENT, IMPROVING QUALITY OF LIFE, AND HABIT CHANGE

Case I—*Pepsodent*

Empirically, an expanding market for a brand is a result of a combination of effective market development and energetic marketing. In about the mid-1970s, the market for toothpaste was very small. The yearly sales of *Pepsodent*, Unilever's toothpaste, were less than 400 tons. A competitor's brand, *Prodent*, was by far leading the market by a wide margin.

A strategy that worked well for Unilever Indonesia in developing the market for *Pepsodent* was to get involved in community development and to raise the community's quality of life. In the case of *Pepsodent*, education on oral hygiene started as early as the 1970s, in cooperation with the Indonesian Dentist Association (PDGI) and the faculty of dentistry at major universities. Unilever supported these organizations with product supplies (toothpaste and toothbrush), oral hygiene educational materials, and oral care tools to educate communities all over Indonesia. Even now, in an era when mass media and technology easily bring information and brand messages to the farthest part of the country, when people are comparatively aware of oral care, similar and even richer educational programs and collaborations with more institutions are still considered as important to educate the society further. *Pepsodent*, with a long history of making unique connections with its consumers through community education and improving their lifestyle, possesses an advantage over its competition, and has been leading the market for three decades. This cooperation with PDGI is still in existence today, and both Unilever and PDGI continuously update this solid collaboration with stronger commitment.

All these practices were categorized as an approach to community and capacity building. Further cooperation was also encouraged to give

FIGURE CS1.2 *Pepsodent* volume (1969–91)

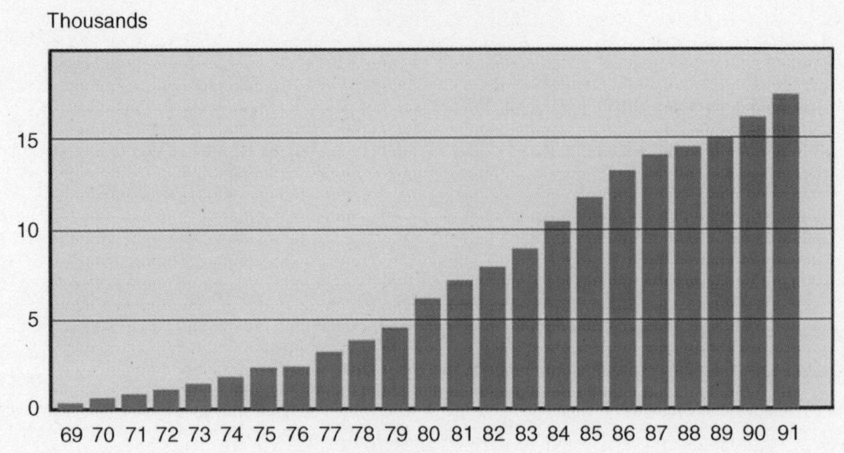

Source: PT Unilever Indonesia, Tbk

free access to dental care by deploying mobile dental units (*Pepsodent* Dental Care Units) operating at schools and suburban and rural areas. Radio, leaflets, and posters were also used to communicate the activities and to raise awareness of oral hygiene and at the same time to raise awareness of *Pepsodent* toothpaste.

The result was that awareness of oral care improved dramatically, especially in rural areas. This improved the quality of life of the population. Unilever Indonesia had forged goodwill among all dental communities, and managed to build equity for the brand, positioning *Pepsodent* as the leader in oral care (see figure CS1.2).

By 1978, *Pepsodent* was positioned as market leader in Indonesia, and this remains the case until today. Unilever Indonesia and *Pepsodent* are able to maintain market leadership, while the community continues to get their oral care education and free dental treatment. The dental professionals and faculties are able to train and develop their students with real practice at schools and rural areas, all financed by Unilever Indonesia.

B. CORPORATE IMPACT ON SOCIETY THROUGH THE DEVELOPMENT OF THE VALUE CHAIN, CREATION OF JOB OPPORTUNITIES, AND WEALTH CREATION (SME DEVELOPMENT PROGRAMS)

Unilever Indonesia has been instrumental in starting up several business partners by sharing the necessary technical know-how and quality assurance, management skills, and good governance operation, and, most importantly, ensuring a captive market. Unilever Indonesia has given a start to several local suppliers to start producing raw or packing materials to substitute for imports. A few examples are:

- PT Sorini, which was founded, in 1983 supplies Unilever with sorbitol, a major ingredient for toothpaste, produced from cassava grown by local farmers. Currently, Sorini is one of the biggest sorbitol suppliers in the world and is exporting to about 55 countries.
- Dai Nippon (PT Dai Nippon Printing Indonesia) is a printing and product-packaging company founded in 1972. Earlier Dai Nippon was a dedicated packaging supplier for Unilever, but over time the company has come to supply a wide range of other clients.
- PT Berlina has produced locally the first laminate tubes for Unilever products such as *Pepsodent.*

As well as suppliers of raw and packaging materials, Unilever has suppliers of services, which mainly are distributors of Unilever products. Unilever helped most of its distributors start up their businesses, giving them continuous guidance along the way, particularly in training their sales promoters. Unilever also defines demarcated area allocations to ensure sustainability and fair competition amongst the distributors.

Internationally Unilever has been working with four large international advertising agencies, which were J. Walter Thompson, Lintas Ltd., McCann-Erickson, and Ogilvy & Mather. Encouraged by Unilever's commitment to growth in Indonesia, all had set up local representative offices jointly with Indonesian companies (the Indonesian law at that time, required any service company entering Indonesia to partner with a local SME). This was an excellent arrangement because these local and international agencies provided more flexibility in handling Unilever advertising matters in Indonesia, which proved to be effective.

TABLE CS1.1 Number of people working full time for Unilever Indonesia within its network

Year	Direct employees	Network
1998	2,300	13,700
2003	3,096	25,000
2008	3,627	300,000

Finally, the third category of business partners is third-party manufacturers and co-packers. With these partnerships, the rules are that the local parties bring in land, buildings, and labor, while Unilever provides the machinery, training, and technical support at the initial stage of the development. The technology of choice is usually simple manual machines requiring greater labor.

It is apparent to Unilever that the development of SMEs within a company's value chain is a type of cooperation with shared benefits. Referring to the figure 3.8, the development of business partners within the company's extended supply chain—from suppliers through production up to distribution of products in the marketplace—through sharing operation excellence and good business ethics resulted in a win–win partnership.

The economic impact to society was enormous, because it generates new jobs and massive employment opportunities. Good and profitable SMEs are agents of wealth creation and good governance. At the same time, SMEs provide reliable sourcing and service facilities for the company, creating operating excellence and competitive edge.

Estimated employment generated by partnerships that Unilever Indonesia has established with its suppliers, distributors, co-packers, and warehousing firms are shown in table CS1.1.

Partnership development with SMEs in Unilever Indonesia, through training, development, and capacity building, also through sharing best practice and transfer of technology, can be illustrated as in figure CS1.3.

Case II—*Sunsilk* Sachet

Similar to the strategy applied for *Pepsodent*, Unilever started with education to develop the market for its shampoo brand, *Sunsilk*.

FIGURE CS1.3 Unilever's partnership development with SMEs in Indonesia

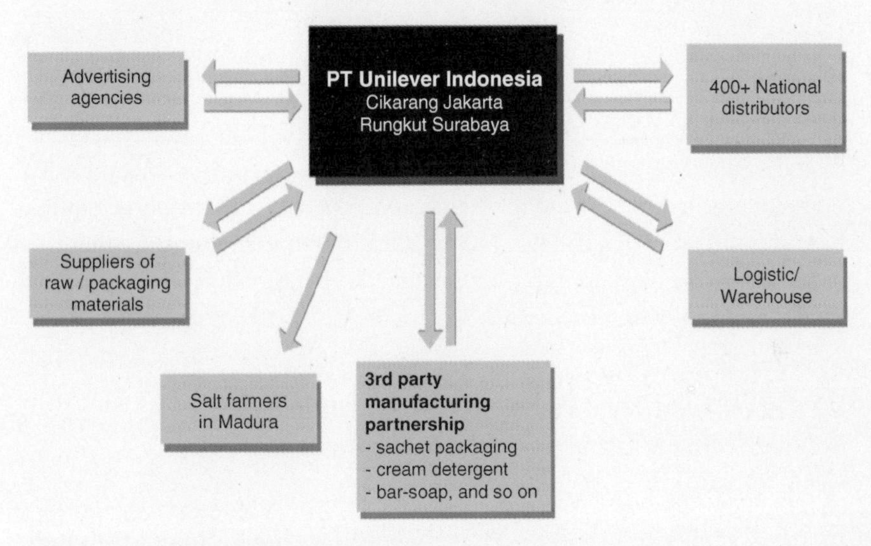

Source: PT Unilever Indonesia, Tbk

Community education to get people to wash their hair using shampoo would have failed to increase the sales of *Sunsilk* had Unilever Indonesia not realized the need to introduce small low-priced packs (sachets) to cater for first-time users, the low-income group, and rural consumers.

Two possible alternatives were available at the time to start the production of the small packs (sachets). The first was to buy high-speed automatic filling machines and produce the sachets in Unilever's own factory premises. This was not considered a reasonable option because the numbers of users of shampoo in suburban and rural areas were still very low, and it was very difficult to project the size of the volume growth. With an initial low capacity usage of high-speed packaging machines counted against the high depreciation cost, the pricing of shampoo per sachet would be distorted and would become too expensive for the target group.

The most effective way was therefore to invest in semiautomatic or manual machines, and move production to a third party. Production capacity could gradually be increased as volume demand grew. In early 1980, four semiautomatic filling machines were transferred to CV Harapan, a third-party co-packer and a small local enterprise. Production of shampoo liquid continued to be done in the factory, and CV Harapan prepared the small packages, did the filling, and arranged all the sachets in display hangers. Unilever provided complete technical guidance and training to ensure quality assurance and production effectiveness, and this decision has enabled CV Harapan to become a full-fledged co-packer company. From a single-person family business operating in a small garage, CV Harapan is now giving employment to more than 100 workers with turnover of IDR3.6 billion (2007) led by skilled and motivated owners.

Case III—Develop and Establish a Sales Distribution Network Through Capacity Building and Training (1980 onward)

In 1989, Unilever Indonesia directly employed more than 100 salespeople and relied on 200 distributors for product distribution. Most distributors were inactive and completely dependent on Unilever Indonesia's salespeople, merely acting as stock points, waiting passively for customers to come and buy Unilever products.

Companies started to invest in Indonesia with many international fast-moving consumer goods (FMCG), and the market became highly competitive. Unilever Indonesia set a new strategy, ensuring an effective distribution network to bring products to consumers. Considering the size of the country, Unilever Indonesia established this network through partnerships with a wide range of independent distributors and subdistributors located all over Indonesia, covering important trading areas. Capacity-building programs were implemented to grow these partnerships. Unilever helped the distributors to become skilled business owners and motivated entrepreneurs, providing training on merchandising, selling, and marketing, as well as warehousing management, stock control, FIFO (First In First Out), and bookkeeping.

Each of the distributors made a three-percent margin on a turnover that ranges from about IDR11 million to IDR440 million, with

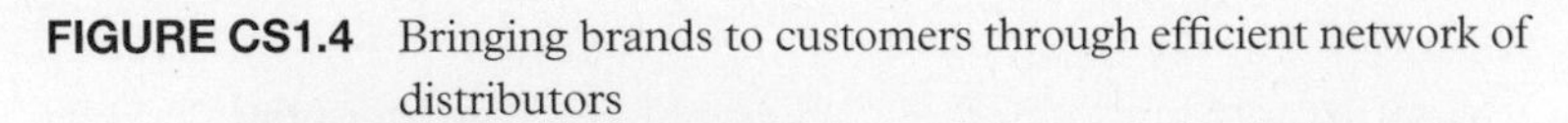

FIGURE CS1.4 Bringing brands to customers through efficient network of distributors

additional bonuses for outstanding sales performance. Sales areas were demarcated to ensure the sustainable profit and growth of each distributor. In 2007, more than 400 distributors covering 750,000 retailers throughout Indonesia employed more than three times the number of people directly employed by Unilever Indonesia (see figure CS1.4).

Case IV—Sourcing of Black Soybeans

In 2001, Unilever Indonesia acquired *Kecap Bango*, a sweet soy sauce producer based in Central Java. Sweet soy sauce is basically made from two agricultural products sourced locally: black soybeans and coconut sugar. Three years after the acquisition, *Kecap Bango* sales increased by

425 percent. With this dramatic increase in demand, it was difficult to get sufficient supply and consistent quality of soybeans purchased from local traders.

Unilever Indonesia therefore started to develop a program to create an alternative supply chain for a black soybean crop sourced directly from local farmers. Unilever Indonesia's involvement in the production of this agricultural raw material was required at this point to secure future demand and guarantee consistent soybean quality because, traditionally, black soybeans are planted and traded through a labyrinthine supply chain of farmers and brokers.

Realizing that the company had no expertise in farming black soybeans, Unilever Indonesia requested the help of the experts from the agricultural faculty of Gadjah Mada University in an agricultural development program. Under this program, the team from the Gadjah Mada University worked to develop a new variety of black soybean seed, improve the quality of the seed, develop certified seed sources, and identify more reliable production methods, while Unilever provided financial credit and guaranteed purchase of the product at a contracted price. The project started in 2002, and in 2008 around 5,000 farmers were participating in the planting of black soybean, covering an area of about 5,000 acres in Yogyakara and East, West, and Central Java (see figures CS1.5 and CS1.6). This project will be expanded to other areas in Java, and also in Bali. Farmers' participation in this project is expected to grow further at a rate of 50 percent per annum.

The outcome of this CSR program gives Unilever Indonesia a secure supply of black soybean with consistent quality. Farmers are able to increase their income with the security of captive market, a better offer because direct purchase by Unilever Indonesia gives them a price of 10–15 percent higher than the market rate. Furthermore, they have the availability of credit from Unilever Indonesia and the university's technical assistance. For the first time in 61 years, Gadjah Mada University's Department of Agriculture became the founder of a completely new variant of premium quality black soybean called "Mallika," certified by the Minister of Agriculture through a decree issued on February 2007.

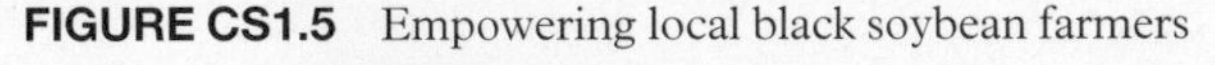

FIGURE CS1.5 Empowering local black soybean farmers

C. GAINING COMMUNITY ACCEPTANCE AND GOODWILL THROUGH CARE FOR THE COMMUNITY AND ENVIRONMENT

Case V—Unilever's Environmental Care Program

The Unilever Environmental Care Program started in 2001 in Surabaya, the capital city of East Java, where Unilever Indonesia operates two factories. While the factories are equipped with an environmentally friendly waste processing unit, Unilever recognized the need to extend the environmental concerns and go beyond its own operations. In partnership with Surabaya City Government and a local mass media group *Java Post*, Unilever Indonesia launched a program not only to improve the

FIGURE CS1.6 Improved soybean seeds for high-quality production

physical condition of the environment, but also to change the community's habit in littering the environment and polluting the river and to mobilize social capital to create a better environment as an ongoing process. This is what is now commonly called sustainable development.

With a vision to improve community life by creating green, clean, and healthy residential surroundings, Unilever Indonesia first began with changing the paradigm of the local people. From just producing waste, they were expected to also manage it with the 4R principle: reduce, reuse, recycle, and replace; and ultimately reach the ideal of a zero-waste environment.

To socialize the program, formal leaders from the highest-ranking executive, the city mayor, to the local district chiefs: Bupati, Camat, Lurah, and neighborhood heads (RT, RW) were encouraged to participate. Likewise with informal public figures, NGOs, and educational institutions. Unilever Indonesia recruited and trained a group of

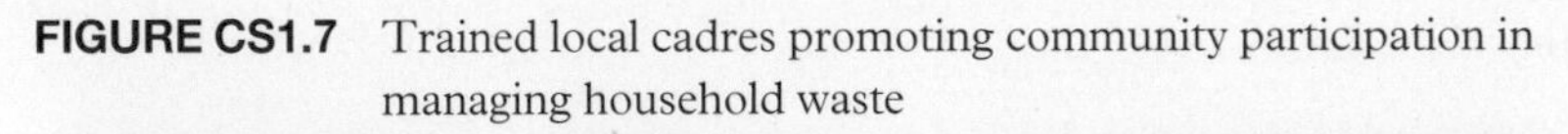

FIGURE CS1.7 Trained local cadres promoting community participation in managing household waste

workers assigned as Unilever's field team to function as motivators. These motivators represented the eyes and ears of Unilever, monitoring and reporting progress in the field. Unilever Indonesia provided the motivators with ongoing training, making sure that their capabilities continued to improve. Results and problems were discussed in routine meetings, and final evaluations were done annually.

Because the key to community involvement in waste management begins at home, the first and foremost task of the motivators in implementing the program was to recruit local cadres selected from informal public figures to act as facilitators to reach these intimate quarters (see figure CS1.7). The local cadres were trained and motivated to become "role models" or agents of social change, promote community participation, and mobilize resources. Motivators trained facilitators in a two-day educational program and equipped them with:

- personal development, covering skills to influence others to take more care of the environment, leadership skills, ways to help people, and presentation techniques

- basic education on the environment and waste selection technique and management
- information on the needs of their particular areas

The training and exercises made the cadres confident in speaking to and mingling with their neighbors and social groups, to educate and bring change to their community. They organized consultancy meetings, visited houses, and participated in all routine community activities. To ensure continuity of the program, they also organized competitions such as "My Clean Village Contest," "The Recycling Dry Waste Contest," and "The River Flood Plan Award." Once a month, facilitators gathered to discuss their capabilities and problems, and also to learn about their progress, the newest environmental issues, techniques, and knowledge.

First accompanied by the motivators, the facilitators started with a small-scale plan, that is, household waste sorting. Each and every house was encouraged to separate three types of waste: wet waste (easily degraded and decayed waste such as food debris, vegetables, fruits, kitchen waste, and garden waste); dry waste (nondegradable waste such as paper, carton box, plastic, cloth, rubber, and tin cans); and hazardous and poisonous materials (batteries, paints, pesticides, and so on), and collected them in three separate bins. Because wet and dry waste has special economic value, facilitators also provided the community with waste-processing skills. Wet waste was processed into compost, which was very useful as an organic fertilizer. This is a simple process and can be done within each household anywhere. Dry waste, such as plastic packaging and carton or paper, was collected and sold to recycling industries. Nonrecycled plastic waste was then converted creatively to useful products such as bags, mats, and umbrellas. Several entrepreneurs in plastic recycling industries have been able to convert the nondegradable plastic into raw material for polyvinyl chloride (PVC) and aluminum sheets for laminated roofs and doors.

From a simple act of waste sorting, motivators and facilitators were able to expand the program to include environmental care and cleanliness, as well as tree planting, in the now motivated society. This is indeed a long process, but five years after the program was launched,

FIGURE CS1.8 Collecting waste

significant improvement in the quality of life of the community, as well as the Surabaya neighborhood environment was evident:

- The Unilever Environmental Program has changed the paradigm of the community. They succeeded in creating a better communal way of life (see figure CS1.8). The narrow lanes are now neat and tidy, decorated with drawings about care for the environment, which are not just slogans, but really a reflection of their active participation and contribution to the community (see figure CS1.9).
- Between the simple unfenced rows of houses with no yards, medicinal plants and potted flowers bring greenery to the neighborhood (see figure CS1.10).
- The clean river is no longer a backyard, but a verandah (see figures CS1.11a–c). Its cleanliness is organized and protected. Children can play safely at every street corner along the riverside.
- The program has been able to increase the individual and community potential in producing innovative and meaningful products,

FIGURE CS1.9 No more scattering rubbish around this now clean neighborhood

theoretically from waste, that subsequently generates opportunities for new sources of income.

The next action would be to focus the program on one area with step-by-step execution to ensure a good learning period before expanding to other areas. From an idea to promote healthy and green living space through community education, this program in Surabaya has been able to change the community habits in dealing with environmental issues. It also improved community lifestyle and provided opportunities for wealth creation (see figures CS1.12a–b). Cooperation with the local government, a local NGO, a local university, and local media was an effective approach for community acceptance and goodwill. Companies that correctly meet the local needs would no doubt sustain a good corporate image.

The effort in developing sustainable environment together with the Surabaya residents was recognized regionally and globally. The program and the city won the following awards:

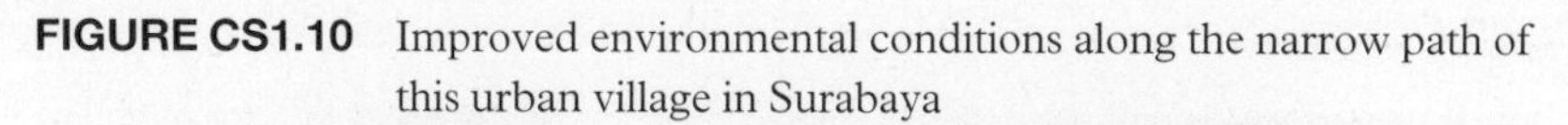

FIGURE CS1.10 Improved environmental conditions along the narrow path of this urban village in Surabaya

- For the first time in 2006, Surabaya received the national annual environmental award *Adipura*.
- The Energy Globe Award for water, judged by the successful implementation of a sustainable program to improve the quality of the environment, particularly around the Brantas River in Surabaya.
- An award and financial gift of €10,000 were presented to Unilever Indonesia at the Global Technology Fair in Vancouver, Canada in 2006.

FIGURE CS1.11A The river before the program

© PT Unilever Indonesia, Tbk

FIGURE CS1.11B People washing clothes in the river before the program

© PT Unilever Indonesia, Tbk

FIGURE CS1.11C The river after the program

FIGURE CS1.12A Products created from waste

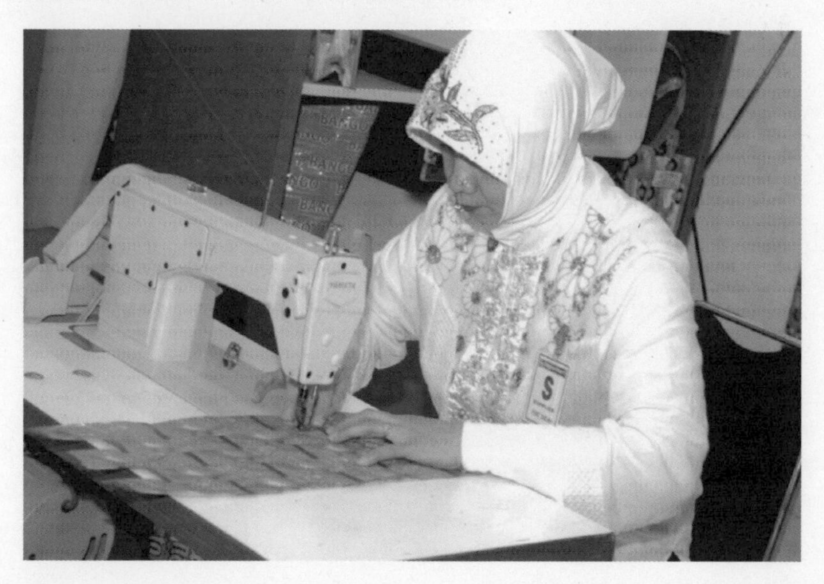

FIGURE CS1.12B The Litterbug Program, promoting female entrepreneurship

The success of this Surabaya Green and Clean Program was expanded to other areas with the same program tag, that is, the Jakarta Green and Clean Program in 2007 and the Yogyakarta Green and Clean Program in 2008. The important lesson here is that by emulating a proven program, the implementation of the similar program in Jakarta is generating results faster.

NOTES

1. With two parent companies, Unilever N.V in the Netherlands and Unilever PLC in England and Wales, and their group of companies.
2. Which, from August 2007, was wholly owned by Unilever.
3. As reported in Unilever Indonesia Annual Report 2007.

Case Study 2

PT BANK DANAMON INDONESIA, Tbk

THE COMPANY

Established in 1956, PT Bank Danamon Indonesia, Tbk (Danamon) is the second-largest private national bank and the fifth-largest commercial bank in Indonesia, with a 5-percent share of the domestic system loans and deposits. In 1988, Danamon became a foreign exchange bank and in 1989 was publicly listed in the Jakarta Stock Exchange.

Following the major regional financial crisis, the bank was recapitalized with IDR32 trillion worth of government bonds. Furthermore, eight banks earlier taken over by the government were merged into Bank Danamon's operation. Subsequent restructuring efforts in every area of the business enabled the bank to establish new foundations and infrastructure to pursue growth based on transparency, responsibility, integrity, and professionalism (TRIP).

In 2003, Danamon was acquired by Asia Financial (Indonesia) Pte. Ltd., through a consortium of Temasek Holdings and Deutsche Bank AG, and took a majority controlling stake. A new vision was introduced, and a universal banking approach was developed with specific business models serving clearly defined market segments.

Through the acquisition of Adira Finance in 2004—one of the largest consumer automotive financing companies in Indonesia—Danamon diversified its business into consumer financing and launched a community banking program it called self-employed mass market, generally known as Danamon Simpan Pinjam (DSP). This program was designed to serve micro- and small enterprises and help them grow. In the meantime, the acquisition of American Express card business in Indonesia in 2006 put the bank as one of the largest card issuers in the country.

Currently, Danamon is recognized as Indonesia's leading small and medium-sized enterprise (SME) and consumer bank with the highest net interest margins. Bank Danamon maintains steadfast commitment to the corporate values: caring, honesty, passion to excel, teamwork, and disciplined professionalism to support the company vision.

CORPORATE SOCIAL RESPONSIBILITY

To increase the impact of Danamon's corporate social responsibility (CSR) programs, in 2006 PT Bank Danamon Indonesia, Tbk and PT Adira Dinamika Multi Finance, Tbk (Adira Finance) established Danamon Peduli Foundation (DPF). It is a fully independent entity that supports sustainable community driven activities and volunteerism. The foundation implements transparent financial accountability through an annual financial audit done by a certified public accountant (good corporate governance).

Through Danamon Peduli programs, more than 40,000 employees working in more than 1,400 Danamon's branches all over Indonesia can participate in contributing toward bringing significant impact to the improvement of the quality of life of their communities through their actions.

DEVELOPING AND EXPANDING THE MARKET THROUGH COMMUNITY BUILDING, IMPROVING QUALITY OF LIFE, AND WEALTH CREATION

Case I—My Clean, Healthy, and Prosperous Market

Danamon Peduli Foundation launched a CSR program called My Clean, Healthy, and Prosperous Market, aimed at improving the cleanliness, hygiene, and performance quality of traditional markets, as well as capacity building of SMEs throughout 1,500 traditional markets in Indonesia. This unique win–win program successfully conditioned the marketplace, at the same time creating sustainable development of the marketplace community, which eventually improved lifestyle and generated wealth. This expanded the market of DSP products.

Innovation

DSP is an innovative product, specially designed to serve the microenterprises and individual entrepreneurs who have no access to the banking system, of which 60 percent are the traditional market vendors. Danamon defines a micro- and small enterprise as any business with an annual selling performance less than IDR2 billion or a business with credit needs of IDR1 million–IDR500 million on average.

In a survey by the bank of 1,000 microbusiness owners in eight cities, 94 percent of the respondents said they need loans to help them grow, but only 36 percent have access to commercial bank credits. They normally borrowed money from friends, family, individual money lenders, or cooperatives. The long process and complicated requirements also prevent them from going to the bank, aside from the fact that these microbusinesses are usually one-man-band organizations, whose owners must attend to their shops and cannot afford to leave their customers to see the bankers. Most of the respondents also perceived dealing with banks as a frightening experience. They need simple services, an easy and fast process, and if possible, transactions done at their business location. Based on this research, Danamon established DSP with innovative products, processes, branch locations, and services specially designed to cater the needs of micro- and small entrepreneurs. They were even involved in the design of the name and logo.

Traditional markets are one of the pillars of the economy and welfare of the Indonesian people, a beehive of activity, and they are always dirty. Especially at grassroot level, traditional markets are central to the socio- and microeconomic development of a big portion of the population. With more than 12,500,000 vendors gathering inside more than 13,000 markets, traditional market communities and their families can reach more than 50 million people. Over the past few years, the traditional markets have been facing severe competition from the modern trade—supermarkets and mini markets—penetrating the local economy. There was no other way to survive, except to strengthen the traditional market's competitive edge by improving its cleanliness, hygiene, and convenience. Clean and healthy markets attract customers, which in turn enable the markets, and thus the community, to prosper. Commitment to the cooperation of all stakeholders, that is, vendors, marketplace management, related government institutions, local non-governmental organizations (NGOs), and DSP volunteers and employees, as members of the traditional market community, is critical to maintaining the existence of traditional markets.

Today, with 700 DSP branches covering approximately 1,500 traditional markets, Danamon Peduli volunteers who are DSP employees are able to work with traders, market managers, local NGOs, and local government to clean and improve community market's physical performance.

Complementing the market cleaning project, the following initiatives are being implemented to support My Clean, Health, and Prosperous Market program:

- small-scale infrastructure development, based on the urgent needs of the traditional market community: public toilets, road access, drainage system, information boards, zoning system, fire extinguishers, and so on, which upgrades market conditions
- donation of fogging machines to prevent a dengue fever epidemic and hand-washing facilities to reduce the risk of diarrhea and avian flu. Another program with the same benefit is the regular competition to build awareness and excitement of all stakeholders in improving the quality of market environment
- community lifestyles are expected to improve with continuous health-related education and activities, such as health campaigns, free medical checkups and blood donation
- DSP to secure its competitive edge with the foundation conducting training of simple financial management for the self employed and small-scale vendors

Launched in 2004, My Clean, Healthy, and Prosperous Market has grown from 730 activities to 1,111 in 2007, involving 11,194 volunteers and touching 558,785 beneficiaries. DSP loan growth in 2007 was 48 percent with very low nonperforming loans. While previously the program was sporadically applied, My Clean, Healthy, and Prosperous Market was conducted simultaneously in 717 markets in 2008 across Indonesia to symbolize the announcement of National Clean Market Day on July 19, 2008 by Danamon Peduli in cooperation with the Ministry of Trade.

A survey conducted by Danamon Peduli started in 2008 among 717 DSP branches/markets to measure the impact of My Clean, Healthy, and Prosperous Market program on multi stakeholders in the market indicates benefits realized by local communities and the related public offices, which are the traditional market management and its cleaning services.

Benefits for The Community

- The market becoming a pleasant and clean environment to work in and trade their goods (42.3 percent; see figure CS2.1)
- Empowering community in contributing and participating actively in keeping the market clean (17.4 percent)
- Optimizing the use of clean and hygienic public facilities (14.0 percent)
- Familiarizing the community to deal with the local DSP branch (12.8 percent)
- The availability of free health treatment (12.8 percent)

Benefits for The Local Government

This particular Danamon CSR activity has supported the local government in the following areas:

- improving the infrastructure of the market (52 percent)
- achieving its social mission (15.9 percent)
- educating the community on the importance of cleanliness and hygiene (5.3 percent)

FIGURE CS2.1 Pasar Beringharjo in the city of Yogyakarta, one of the main traditional markets participating in the My Clean, Healthy, and Prosperous Market program

- Strengthening the role of the local government to the community (3.7 percent)
- developing and establishing good cooperation and relations between the DSP branches and each local government (17.3 percent)

Case II—Danamon Go Green

The traditional markets in Indonesia produced thousands of tons of waste every day, of which 70–90 percent is high-quality materials for organic fertilizers.

In cooperation with local governments and a prominent biotechnology institute, Danamon Peduli applies a simple technology to convert thousands of tons of organic waste into high-quality organic fertilizer. This integrated waste management program improves health and hygiene conditions, as well as generates socioeconomic benefits for the community. Every day, one composing unit can convert one-to-three tons of organic waste into 1,000 pounds to 1.2 tons of compost. Each unit is operated by three or four workers. The whole program provides additional income for the community, solves fertilizer scarcity issues, improves land condition, reduces the waste management problem, and so on.

The key success and sustainability of the Go Green program is a win–win partnership between DPF and the local governments. Danamon Peduli provides the project design, machinery and composting house, technical assistance, training, monitoring, and evaluation. The local governments identify needs, provide land within or close by the markets, manage local coordination, issue permission, and socialize the program. To ensure management commitment, the program is included in the local government strategic planning. Danamon Peduli rolls out the program all over Indonesia, while local governments replicate the donated units in other traditional markets within their region. In the year 2008, 31 local governments agreed to replicate the program and by the end of the year (2008), 26 local government had already signed a memorandum of understanding with DPF.

A successful example on the usage of organic fertilizer is in Bantul–Yogyakarta, where the organic fertilizer has increased the shallot harvest by 30 percent in 42 acres of sandy coastal area, and has reduced the use of chemical fertilizer by 70 percent (see figure CS2.2).

The first national convention was held in Bantul–Yogyakarta on December 17–18, 2008 with a theme "Market waste organic fertilizer to improve prosperity of traditional markets and national food resilience based on organic farming," in which an agreement has been reached on standards indicators to measure the success and sustainability of the program.

The local governments send a progress report every month to be processed by Danamon Peduli knowledge management system. Through this monthly reporting system, each local government get a regular feedback on their result compared to others.

Note: This Danamon Peduli CSR program "Nothing wasted—Indonesia" was selected as the second winner of the BBC World Challenge 2009 award at The Hague on December 1, 2009

Case III—Philanthropy

Unlike the market programs, which deal with sustainable development of the community, the third program, called 3R: Relief, Recovery,

FIGURE CS2.2 The conceptual design of the traditional market waste management system in Bantul, Yogyakarta

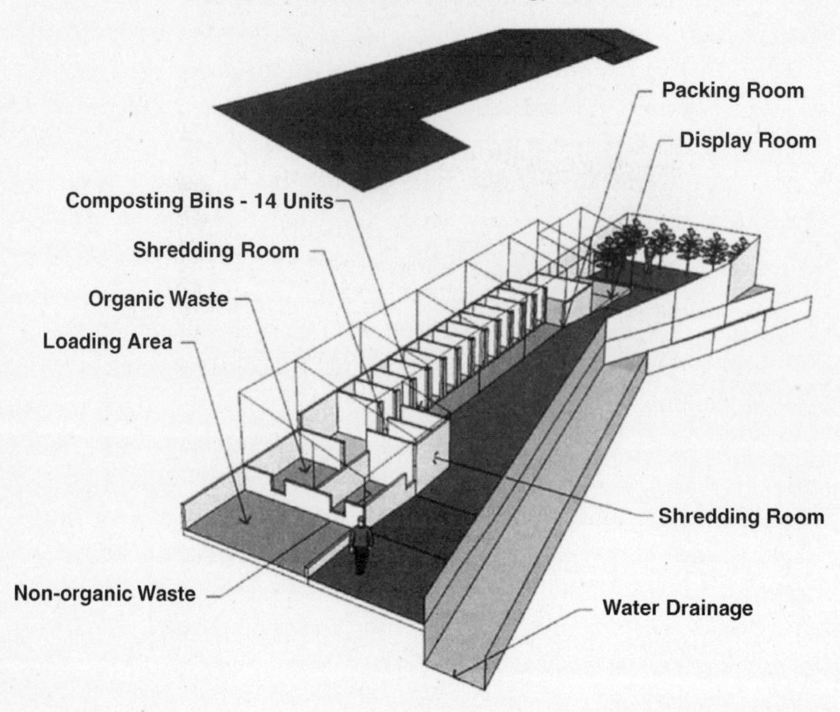

Reconstruction, concentrates more on active humanitarian aid, providing support to the victims of natural disasters and help them develop and recover their livelihood.

Indonesia is located in the Pacific Ring of Fire, so many of the provinces are vulnerable to natural disasters, particularly earthquake, tsunami, floods, and volcanic eruptions.

Danamon strives to be the first private corporation to help victims of natural disasters in Indonesia. Danamon Peduli has developed a simple mechanism for all Danamon branches to get quick approval for funds through a text-messaging system that enables them to act quickly when disasters occur.

In 2008 Danamon Peduli helped more than 17,000 victims of 37 incidents of disasters in North Sumatra, Jakarta, West Java,

FIGURE CS2.3 Supporting the community in the aftermath of natural disaster

Yogyakarta, East Java, Central Kalimantan, East Kalimantan, South Sulawesi and West Papua, engaging 575 employees (see figure CS2.3).

Case Study 3

PURA GROUP

THE COMPANY

PURA is an integrated group of companies, one of the leaders in Southeast Asia in the industry of papermaking, printing and packaging, paper converting, holography, total security systems for anticounterfeiting, and smart-card technologies and engineering. The company has pioneered numerous innovative products and processes, which include:

- *First in the world:* Hologram on aluminium for blister packs, scratch-off holograms on prepaid phone cards, and the modification of one color K-offset printing machine into two- and four-color Intaglio Printing Machine
- *First in tropical countries:* No carbon required (NCR)/carbonless paper
- *First in Southeast Asia:* Microcapsules/NCR (main material, security hologram and hot-stamping foil, total security system, private security paper and banknote paper producer, melamine decorative paper, CO_2 laser cigarette tipping paper) perforation machine, and nanotechnology in producing solar window film
- *First in Indonesia:* Voucher card for cellular telephones, smart technology producing contact and contact-less cards

The company was founded in 1970, when the current owner/president director of PURA Group inherited the company from his father. It was then a vulnerable small offset printing company with 35 employees. He directed a new vision and business development toward specialization to achieve and maintain perfection. The underlying philosophy of the company is the commitment to ensure an ongoing innovation and development of high-technology special products, with high local content for import substitution, catering for the local and the global markets. PURA Group has demonstrated capability to adopt technology to suit the environment of a developing country, especially Indonesia.

Under this new leadership, PURA has now secured 73 registered technology patents and, so far, exports its products to more than 94 countries in Asia, the Middle East, Africa, Australia, Europe, and the US. This group of companies—which consists of 24 interrelated production divisions with complementary businesses that use its innovative and adaptive capabilities in the technology areas—directly employs more than 8,500 employees, which also contributes to indirect community employment of more than 40,000 people and their economy's wellbeing.

CORPORATE SOCIAL RESPONSIBILITY

PURA's commitment to the community is reflected in the company's corporate social responsibility (CSR) vision:

- commitment to the development of community values
- use of technology, capacity building, and empowering and creating wealth for the community
- developing and sustaining the growth of small and medium-sized enterprises (SMEs) to ensure acceleration of microeconomic and people-based economy growth

Case I—Extended Supply Chain and Wealth Creation

Being responsive to the needs of the local community is the driving force behind PURA's commitment to microeconomic growth. Empowering the local people will ultimately generate a positive impact for the company. Several CSR-related activities carried out by PURA create new sources of income for the society, and also increase the company's competitive edge. PURA is now one of the leading companies in the comprehensive printing and packaging industry across Southeast Asia.

In paper manufacturing, paper waste is often considered to be a significant raw material, besides pulp. PURA requires about 12,000–15,000 tons of paper waste per month, worth almost IDR15–20 billion per month. In 1975, PURA started an education program for the local community to socialize the benefit of paper waste collection. Currently, paper waste collection is a business on its own for many small companies, and they develop into reliable suppliers for PURA and other local paper manufacturers. They even enter the export market.

To reduce coal consumption and to use more environmentally friendly fuel, PURA has modified burners of its electric power station boiler to run simultaneously on coal and biomass, such as husk and jatropha cake (waste of jatropha oil processing). PURA needs more than 5,000 tons of husk per month, and the supply is fulfilled from purchasing from the community. Husk collection, transport, and unloading are organized and done by small local companies run by the local people. Husk, the waste product of rice milling plants, with very low value, becomes a useful material with high economic value.

Case II—Leveraging Technology to SMEs

Supported by technologically skilled human resources, PURA helps SMEs improve their operational time and efficiency through the creation of simple, affordable, and good-quality equipment. Although PURA may not directly benefit from this activity, the company enjoys healthy social acceptance.

Paddy-Drying Machines

PURA has invented and marketed husk-fueled paddy-drying machines (see figure CS3.1). Husk, which is abundantly available in rice-milling plants, is an ideal substitute for expensive diesel fuel. The use of husk to fuel the machines is not only environmentally friendly, but also drastically reduces the financial burden to the farmers arising from the cost of diesel fuel. The heat generated from husk is almost sufficient to dry the

FIGURE CS3.1 Easy-to-operate paddy dryer, using simple technology

© PURA Group

paddy, not counting the energy required to operate the mechanical parts.

Compact Ice-Flake Machines

These compact machines are now widely used in the fishing industry to improve the operational efficiency of cold storage on fishing boats. Ice flakes preserve fish longer after capture, so fishermen are able to spend more fishing days at sea to increase fish yield, and consequently their income dramatically. With the availability of ice-flake machines, consumers of marine products are also protected from the harmful effects of dangerous chemicals often used by fishermen as substitute preservatives to economize the use of ice for refrigeration purposes.

In comparison with ice blocks, ice flakes have a lower temperature (14–17F) and therefore higher cooling power. Flat, thin ice flakes make better contact with fish and cover all parts without damaging the fish.

Ice-flake machines are compact, flexible to use, provide short production starting time and are of lower investment, with the same capacity over the conventional block ice-making machine.

Ice flakes improve fish quality, and the wider use of the machine will strengthen the competitive edge of the Indonesian fishing industry. If spread nationally and effectively coordinated, Indonesia can potentially play a bigger role in the fish export market. Indonesia has the longest coastline in the world, with vast marine territory, but its fishing exports are only one-quarter that of Thailand ($5–6 billion a year).

Waste-Fueled Biomass Cooking Stove

Many impoverished communities in Indonesia depend heavily on wood to fuel their simple traditional stoves for household cooking. The kerosene stove, although for many years were considered as a cheap alternative, is no longer economical, after the government removed the subsidies on kerosene. The government's conversion program to

FIGURE CS3.2A Specially designed ice-flake machines, for easy use in simple fishing boats

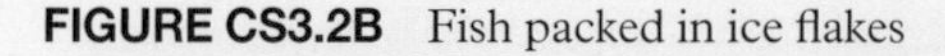
FIGURE CS3.2B Fish packed in ice flakes

© PURA Group

substitute the consumption of kerosene to liquefied petroleum gas (LPG) has been running for almost two years, but many people found the program failed to address the real energy needs of the poor. Besides, the price of LPG is still beyond affordability for the lowest-income group, and there were also cases of lack of safety in the special LPG canisters and stoves distributed to the poor families.

PURA is actively developing an alternative solution to provide the underprivileged with a clean, cheap, effective, renewable, and environmentally friendly fuel cooking stove. Biomass—organic material derived from plants, crop waste, wood waste, or animal waste—is a renewable source of energy for heating and cooking. The traditional method commonly used, namely the direct burning of biomass, creates strong-smelling smoke and soot. The biomass cooking stove, designed by PURA, applies a technology, called forced-draft cyclone, to produce almost smokeless, sootless combustion, due to the perfect burning of biomass in the form of any organic waste: household waste containing

FIGURE CS3.3A AND CS3.3B Prototypes of the PURA biomass cooking stove

(a)

(b)

minimum plastic material, charcoal and noncharcoal briquettes (see figures CS3.3a and CS3.3b). Two pounds of biomass fuel the stove for about one hour, offering significant savings compared to the use of an LPG or kerosene stove (see figures CS3.4a–c).

FIGURES CS3.4A–C Different types of waste: jatropha waste, crop waste, and household waste fueling the stove

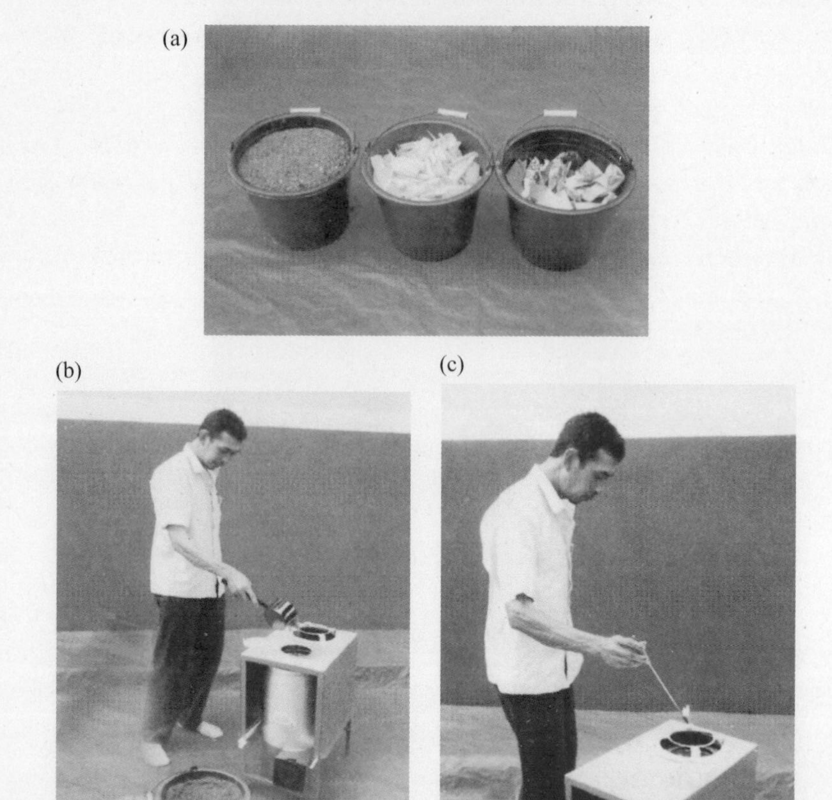

Case III—Capacity Building and Environmental Protection Through the Development of Clean Alternative Energy

This project is designed as part of the rural community empowerment and welfare improvement program, and also to support government policy in the development of clean alternative energy (green fuel) to replace conventional mineral fuel.

The main focus of the development of a *jatropha curcas* (castor plant) as 100 percent pure biofuel is mainly to substitute for diesel fuel and kerosene. *Jatropha curcas* is a tropical plant that can easily grow on any type of ground, even on arid, nonfertile, and unproductive soil. Its cultivation will not disturb the existing land for agriculture or forestry. Unlike cane molasses and palm oil, *jatropha curcas* has no food application for human consumption and therefore it makes an ideal source of renewable energy.

To ensure easy and ongoing production of *jatropha* biofuel as an energy alternative to diesel oil or kerosene, for the past two years PURA has also developed the necessary support in the form of:

- design, preparation, and manufacturing of *jatropha*-processing miniplants to produce biofuel with various capacities for installation in cultivation centers
- a production plan for a manual *jatropha* seed press for the home industry
- a production plan for an affordable biofuel stove
- development of an adapter kit for diesel engines to run on pure biofuel. The cost of the kit is marginal and it can easily be installed without any special technical know-how.

PURA facilitates a training center in Kudus, providing a machinery showroom, a comprehensive classroom and field training, and also a model plantation for *jatropha curcas*.

Gunung Kidul Jatropha Cultivation Pilot Project

PURA has initiated *jatropha curcas* cultivation in Gunung Kidul (a poverty-stricken district on the south coast of Java, in Yogyakarta province). PURA distributed seeds for free to the community and has committed to buy all harvested seeds. In the following years, *jatropha* cultivation will be expanded to Temanggung and Wonogiri (Central Java), Pacitan districts (East Java). Currently, 25 million *jatropha* plants have been cultivated. The number was expected to reach 50 million by the end of 2009. This project could create work and business opportunities for about 1.5 million people.

Mainly, the *jatropha* seeds are processed into biofuel. Meanwhile the seed cake, the waste of seed pressing, has heat content (about 21 kilojoules) that is equal to good-quality coal and can be used as an alternative fuel to replace coal in power plants. The collection and transportation of seed cakes present another business prospect for many small enterprises.

Starting with Gunung Kidul, this program is expected to become a pilot project in converting poverty into prosperity.

Case Study 4

PT ASTRA INTERNATIONAL, Tbk

THE COMPANY

Astra is a public company listed on the Indonesian Stock Exchange, with market capitalization of IDR110.5 trillion ($10 billion) in 2007, and more than 100,000 employees. The company was established in 1957, initially registered as a general trading company, particularly involved in agricultural trading. Twelve years down the road, in 1969, Astra was granted sole distributorship for Toyota vehicles in Indonesia, and Astra has now developed into one of the largest diversified business groups in Indonesia covering the business areas of automotive, financial services, heavy equipment, agribusiness, information technology, and infrastructure (see figure CS4.1).

CORPORATE GOVERNANCE AND CORPORATE SOCIAL RESPONSIBILITY

Corporate governance is already a way of life at Astra because it has been an integral practice within the company for more than three decades. Astra believes that corporate governance balances the needs of different stakeholders, and helps build a successful and sustainable business. By applying a high degree of control within its operations for such a long time, Astra has been able to maintain community-related

FIGURE CS4.1 Business divisions in PT Astra International, Tbk the largest diversified business group in Indonesia

Automotive	**Automobile** • Toyota • Isuzu • Daihatsu • BMW • Peugeot • Nissan Diesel (Truck)	**Motorcycle** • PT Astra Honda Motor	**Components** • PT Astra Otoparts, Tbk	**Others** • AstraWorld • PT Serasi Autoraya	
Financial Services	**Automobile Financing** • Astra Credit Companies • PT Toyota Astra Financial Services	**Motorcycle Financing** • PT Federal International Finance	**Heavy Equipment Financing** • PT Komatsu Astra Finance • PT Surya Artha Nusantara Finance	**Banking** • PT Bank Permata, Tbk	**General Insurance** • PT Asuransi Astra Buana
Heavy Equipment		**Construction Machinery** • PT United Tractors, Tbk • PT Traktor Nusantara		**Mining Contractor** • PT Pamapersada Nusantara	
Agribusiness		**Crude Palm Oil** • PT Astra Agro Lestari, Tbk			
Information Technology		**Document Solution** • PT Astra Graphia, Tbk		**IT Solution** • PT SCS Astragraphia Technologies	
Infrastructure		**General Infrastructure** • PT Astratel Nusantara • PT Intertel Nusaperdana • PT Marga Mandala Sakti • PT PAM Lyonnaise Jaya			

Source: PT Astra International, Tbk

programs that truly benefit the society and the nation, at the same time giving Astra a distinct and sustainable competitive advantage.

Social, community, and environment-related activities, which are now categorized under corporate social responsibility (CSR) programs, have long been implemented as an integral part of Astra business as a means to achieve economic, social, and environmental balance. In 1980, the founder of the company started with the development of small enterprises, by establishing a foundation to help potentially competent small and medium-sized enterprises (SMEs), within the business supply chain, to grow and operate competitively in the national industrial structure.

The formation of this foundation was the initial step toward widening Astra's CSR commitment to become a socially responsible and environmentally friendly corporation. In November 2007, Astra released the *Astra Friendly Company Instruction Book*, as the standard reference for CSR implementation.

Astra has established a clear structure for its CSR programs which was derived from the corporate philosophy (*Catur Darma*, The Four Good Deeds) and the company vision (see figure CS4.2). The programs were designed to be proactive, structured, and sustainable. Being part of the company strategy, the programs run parallel with the operational business requirements. The implementation of the programs starts with the groups closest to the company, that is to

FIGURE CS4.2 Astra's CSR structure

Source: PT Astra International, Tbk

include employees and their families, followed by programs that cover the community around the company's areas of operation, and further to involve and influence the broader national environment and society.

All these activities are developed based on the CSR implementation framework, which is divided into two areas: program (referring to the execution within the company and district- and countrywide) and work system. Within the work system, the company adopts two standards: Astra Friendly Company—the touchstone for the social application of CSR; and Astra Green Company, which is specifically related to the environment, health, and work safety.

THE DEVELOPMENT OF CSR PROGRAMS WITHIN THE CORPORATE PHILOSOPHY

Expanding the Market and Winning Competitive Edge Through Supplier Development

Astra has long realized that the role of its suppliers was of vital importance to the company's success, because suppliers are part of the production chain and Astra relies heavily on their performance. This is the reason that Astra formed a foundation (named *Yayasan Dharma Bhakti Astra*—YDBA) in 1980, to develop its suppliers and help them increase their productivity, creating a win–win partnership. The foundation provides training to equip suppliers with the required human resources and technical and management skills, facilitates their financing and marketing efforts, and also helps them in the area of information technology.

Currently, Astra works with more than 1,000 subcontractors and vendors, most of whom are SMEs. YDBA continues to play a major role as part of the Astra Group of companies by continuously giving general assistance to the SMEs (see figure CS4.3).

In choosing the suppliers, Astra uses a set of selection procedures, supported by the implementation of international ISO 9001-2000 standards. Each company within the Astra Group sets clear selection criteria, and all associated sections are involved in the process. However, Astra also implements multisourcing strategies not only to secure the supplies required for the production, but also to provide greater opportunities for SMEs to work together with Astra. The

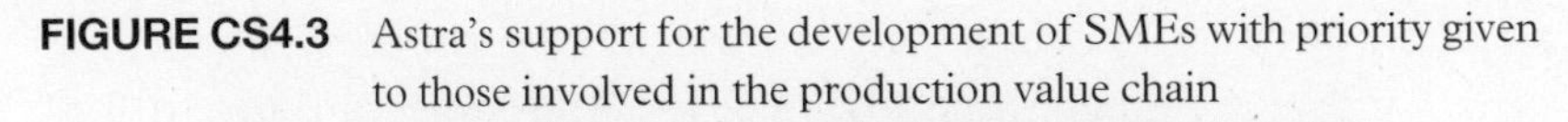

FIGURE CS4.3 Astra's support for the development of SMEs with priority given to those involved in the production value chain

appointed suppliers continuously receive development support, mainly directed at improving the ability to fulfill quality requirements, cost, delivery, and innovation (QCDI). Astra nurtures the partnerships with equality, independence, and transparency as the guiding principles.

In the automotive and heavy equipment business division, the YDBA Foundation, together with the group, offers close assistance, training, and marketing and financing facilities to subcontractors, small repair workshops, and two- and four-wheeler garages, facilitating transfer of technology to help them grow into independent SMEs. The development process prepares the subcontractors for QCDI standards required by Astra, allowing them to start receiving orders. The size of orders passed to subcontractors increases in proportion to the size of the group's overall business growth. Intensive development will last two years, followed by *ad hoc* consultancy sessions. The foundation then begins a new development cycle to give the same competency improvement opportunity to new potential subcontractors.

After the 1998 economic crisis, motorcycles sales in Indonesia increased significantly, from about 14 percent yearly growth before the crisis, to at least 35 percent. Poor public transportation facilities, availability of low-interest credit for motorcycle purchases, and the lack of production and sales regulations for the two-wheeler vehicle contributed to the astonishing multiplication of motorcyclists. The increase in motorcycle sales has stoked a growing demand for motor vehicle parts and services.

While the number of authorized service centers is limited, Astra has seen this as an opportunity for small motorcycle repair workshops or individual mechanics to get involved in the explosion. A big portion of the new motorcycle owners are lower-to-middle-income earners, who are either uncomfortable or cannot afford taking their vehicles to authorized service centers. On the other hand, many small and unauthorized service centers are traditionally managed, and do not have the necessary technical knowledge and tools to cope with the greater need of quality service.

In cooperation with one of its subsidiaries responsible for the sales of Astra genuine parts, PT Astra Komponen Indonesia (ASKI), the YDBA Foundation developed a community-based motor vehicle repair workshop, called *Bengkel Mitra Aspira* (Aspira Partner Workshop). Aspira is the brand name for Astra-guaranteed quality spare parts for motorcycles. The foundation provides technical and management training to selected repair workshops and mechanics, while ASKI helps with design of the workshops and the supply of spare parts. With this type of partnership, Astra can have these small repair and servicing workshops as outlets for Aspira, which enable them to generate profit from spare parts sales.

With the support of several government institutions, state-owned enterprises, and private organizations, the YDBA Foundation also offers training and vocational education for school dropouts. Qualified youths are recruited through a tight selection process to attend training designed to build their knowledge and expertise, and prepare them for jobs as mechanics or small repair workshop entrepreneurs. Encouraging school dropouts to participate in such programs helps these young people to become productive community members.

COMMUNITY AND CAPACITY BUILDING FOR WEALTH CREATION

The YDBA Foundation, since its formation in 1980, is committed to encourage SME growth, inside and outside the company's business line, based on the philosophy of becoming a valuable asset to the nation. With a mission to develop SMEs within the company's value chain, and to empower the local economy by helping diversified SMEs, the foundation works together with each business unit within the Astra Group to ensure effectiveness. YDBA also promotes wider business networks to open up new markets for the SMEs' products.

In line with the CSR implementation blueprint within the Astra Group, the foundation focuses its programs on education and income-generating activities, and based on the scope, divides the programs into four main categories:

- training, consultation, and business support
- management, technology, and marketing development, and financing assistance
- community development around the plantations and mining areas, in cooperation with the Astra Group of companies
- entrepreneurship training for retirees

The fourth type of program is unique compared to the other three. Apart from mere company obligation, this program provides new opportunities for retirees, helping them to keep and increase their economic productivity. The following is a testimonial story of a retiree who successfully became an entrepreneur after his retirement in 1997. Pak Akan Natadihardja, now the owner of a medium-sized authorized service station, has his wife and children helping him to manage his company. The service station posted revenue increase of more than 400 percent in the first 10 years.

Pak Akan and His Service Station

The financial crisis hit the Asian region in 1997, about the same time Pak Akan reached his retirement age, just when he had to leave his job and the only way he knew to earn money. Pak Akan understood that the retirement money would soon disappear with the unreasonable price increase of domestic needs, as well as the swelling education fund for his children. He was left with no choice but to find a new way to support the family.

Opportunities come to those who look for them, and Pak Akan saw the growing number of motorcycles around his hometown as an opportunity knocking on his door. Without a second thought, he agreed to accept the Astra Motorcycle Service Crash Program (ASMD), a program designed to facilitate financial aid and access as a Astra Honda Authorized Service Center (AHASS) offered by YDBA. By October 1997, Pak Akan had sent his work and business plan, complete with the proposed location of his future service station, to YDBA. He also sent two apprentice mechanics to Jakarta to get their first training.

The first few years was a learning period for Pak Akan, just like a baby crawling before he learns to walk. The service center business was a new thing for him, and it was not easy to change his old mindset and to develop new habits as a business owner. YDBA supported Pak Akan and his new company during this phase, helping him with technical and management skills required for him to move the business. To encourage operation efficiency, YDBA also provided the service center with free units of mechanical thruster, bike lift, mechanical tools, air compressor, air duster and piping system.

Now Pak Akan was ready. With a soft loan of IDR8 million (about $800) also provided by YDBA, Pak Akan restored a small leased store and opened his service workshop (see figure CS4.4). During the first month, the service center attracted only a few customers; in three months, the number of motorcycles that came for repair and maintenance amounted to only 10 units; but in one year, the number shot up to 200–250 units per month. Yearly revenue totaled IDR90 million (about $9,000).

Pak Akan exercised his favorite motto, which was "exert your utmost effort in everything you do with enthusiasm." After seven years,

FIGURE CS4.4 Pak Akan's first service station

he moved his service center to a new building, now a building of his own, bought under a low-interest-rate installment plan (see figure CS4.5). This relocation was approved by YDBA, PT Astra Honda Motor, and PT Daya Adira Mustika, another company under Astra Group, as the main dealer, and they also provided new servicing facilities and equipment. Now six mechanics in the workshop were fully trained, ready to offer quality repair service to more than 600 motorcycles each month. The business had grown, with yearly revenue of more IDR400 million (about $40,000) and tangible assets of IDR1.5 billion (about $150,000).

Pak Akan said he owes his success to the synergetic relationship he managed to build with the customers, dealers, and banks through the programs recommended by YDBA.

Astra Green Company: A Framework For Sustainable Business

To achieve the corporate vision of being a socially responsible and environmentally friendly corporation, Astra is committed to the implementation of the Environment, Health, and Safety (EHS) regulations and the Social Responsibility (SR) philosophy. Astra has issued a set of

FIGURE CS4.5 The service station now after relocation

EHS and SR Policy to provide direction for the Astra Group in the application of the EHS and SR system of management. The commitment and direction are also supported by a working framework to ensure improvement in EHS performance. This working framework is the Astra Green Company, designed as a systematic guide to all EHS operations within the group (see figure CS4.6).

Astra Green Company demands green strategy, green process, green product, and green employees, which is a standard that is applicable to all business units. Coherent and harmonious implementation of these four pillars is key to the achievement of company's goal of economic, social, and environmental balance for sustainable business. Astra provides training to ensure all employees, from the top-level executive to the production-line workers, receive proper education and share the same knowledge, thinking, and behavior. In the production process, suppliers are also expected to comply with the green requirements and standards set forth by Astra.

The EHS programs in Astra and its subsidiaries are designed to improve the quality of the environment, as well as tackling the environmental issues and the conservation of natural resources. Astra uses the

FIGURE CS4.6 The concept of Astra Green Company

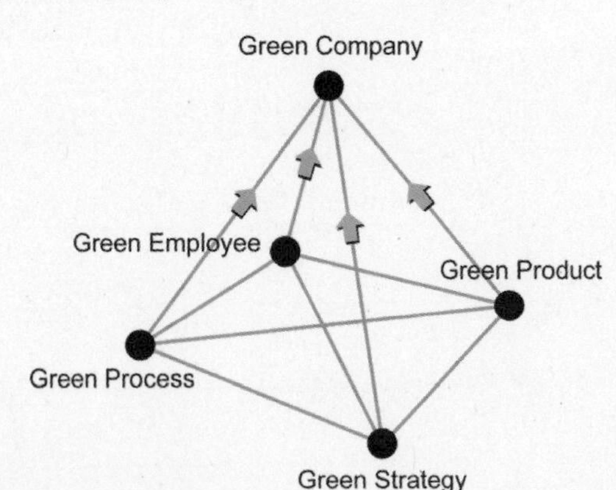

Source: PT Astra International, Tbk

following indicators to measure its ecological footprint: use of natural resources (water, electricity, and diesel) by each production unit, waste (solid and liquid) reduction, number of workplace accidents, and cost and benefit analysis of EHS programs.

The path to sustainable business at Astra was created in 1980, when the early EHS and environmental regulations were first introduced, followed by several development milestones toward a green company along the way. The first initiative concerning the environment was more of a reactive program, and did not help the company to create a sustainable business. In the 1990s era, the company developed a more up-to-date proactive initiative, which was translated into the 4C Approach (commitment of the management to deal with the hazards and pollution from its operations; competence development; compliance to EHS standards; and creating practices of cleaner production in respective work processes). This 4C Approach gave way to the birth of Astra Green Company, and provides the foundation for the Astra Green Company assessment criteria.

Case Study 5

PT INDO TAMBANGRAYA MEGAH, Tbk

THE COMPANY

PT Indo Tambangraya Megah, Tbk (ITM), previously Banpu Coal Operations Indonesia, was established in 1987, and is a subsidiary of Banpu Minerals Thailand Pte. Ltd. In December 2007, the company was listed on the Jakarta Stock Exchange, making ITM an Indonesian publicly listed company. ITM operates four contract areas and two authorization areas with three dedicated ports and ship-loading facilities. The coal-mining operations of ITM are carried out through its five subsidiaries, two of which are in production: PT Indominco Mandiri, PT Trubaindo Coal Mining. PT Kitadin has suspended production in 2006, and restarted in 2009, while PT Bharito Ekatama started its first production in 2009. All subsidiaries are located in East Kalimantan, except Jorong Barutama Greston, which is in South Kalimantan (see figure CS5.1).

The company's operations include coal mining, processing, and logistics. They produce thermal coal with low ash, relatively low sulfur content and with caloric values ranging from 2,400 kilocalories per pound to 3,300 kilocalories per pound. As at December 31, 2007, the estimated coal resources were 1,494 million tons with reserves of 237 million tons. Production in 2007 was 17.7 million tons.

FIGURE CS5.1 Locations of ITM's businesses

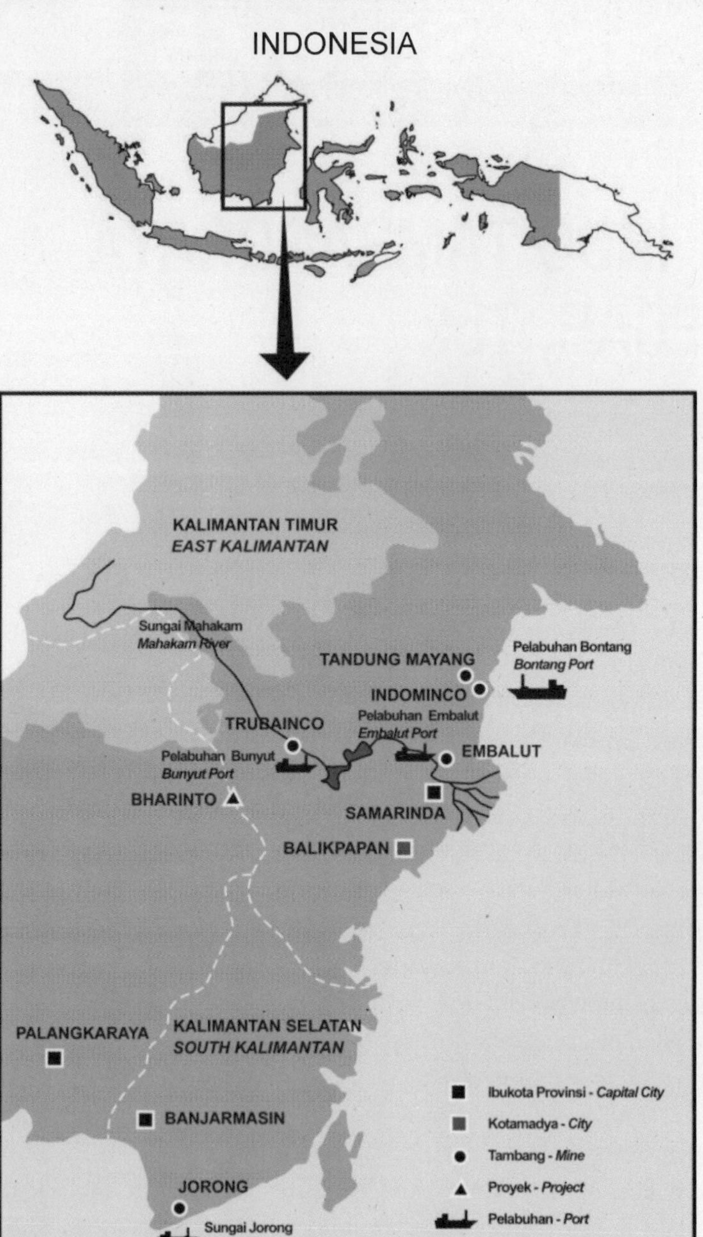

Vision

The company vision is to become a leading coal-related energy company in Indonesia, with sustainable growth through professionalism and care for employees, community, and environment.

Mission

- To develop excellence in all operations, to serve customers with consistent quality and quantity of product and services
- To develop competent employees, efficient systems and infrastructure, under the innovation, integrity, care, and synergy culture of the organization
- To invest in coal-related energy businesses to strengthen the company's position
- To promote and contribute to the development of society by acting as a good citizen and contributing to the economy and society

Corporate Shared Values (Management Guiding Principles)[1]

Being part of the Banpu Group of companies, ITM has adopted the core values and attributes of Banpu's corporate culture, which is known as the "Banpu Spirit." This comprises four main pillars: innovation, integrity, care, and synergy, which have been the foundation of the company's success in the past and will continue to be in the future.

These values are as follows:

Innovation

ITM employees continually improve the quality of everything they do. Through new wisdom, they initiate and are proactive, at the same time they are flexible and adaptive to change. They aim to think "out of the box" and think ahead. They face new challenges with commitment, creativity, and courage to create an innovative culture environment.

Integrity

ITM employees act with integrity, honesty, and a sense of discipline and ethics. They are trustworthy and keep commitments.

Care
ITM employees communicate openly, and are warm and friendly. They care and respect each other, business partners, and all other stakeholders.

Synergy
ITM employees are committed to find the best solution (a win–win solution) for everything they do. They believe in working as a team to achieve sustainable growth and profit and ensure long-term value creation for shareholders. The competitive edge of the company to deal with future challenges depend on the values and professionalism of its people.

Good Corporate Governance

ITM's management has made good corporate governance the culture and standard of the company. The commitment to the principles of transparency, accountability, and compliance with the code of ethics and company procedures have ensured that ITM's shareholders and stakeholders are always prioritized, while ITM continues to operate with responsibility toward the environment and society.[2]

ITM strongly believes in sustainable development and the creation of long-term values for shareholders. As a public company in Indonesia, ITM implements policies to ensure that it maintains effective communication and transparency with all of its stakeholders, investors, and regulators.[3]

Human Resources

ITM has been operating in Indonesia for more than 20 years, and has built its business based on the importance of moral and ethical values. The company puts a high priority on providing training and development programs for its people, as well as making it a pleasant and professional working environment. There are more than 2,500 employees in ITM of different genders, races, religions, and nationalities. The company promotes feedback and equal opportunities for staff at all levels. The staff are dedicated and highly skilled and they work together in various fields. The employees are essentially the soul of ITM.

Quality, Safety, and Environment

Each ITM business unit sets measurable quality, safety, and environment goals, targets, and performance indicators to be compliant with applicable rules and regulations, both internal and external. Corporate goals, targets, and performance indicators and stakeholders' expectations are documented, communicated, monitored, and reviewed for continuous improvement.

Quality Control

Through the quality control system, ITM produces coal that consistently meets and exceeds customers' specifications. The system is carried out using the best practices in achieving the highest standards, effective cost control, and punctual delivery of the end products. Quality control implies the welfare and quality of ITM's employees as well. The company believes in creating a pleasant and professional working environment, while providing a dynamic learning platform for its employees.

Safety

ITM places the highest priority on operational safety and health, which requires strict compliance with rules and regulation to achieve a zero-loss rate for accidents.

Environment

Each business unit of ITM conducts environmental management and monitoring programs, according to the approved environmental impact assessment document (AMDAL) and other related legal and regulation requirements. The company's mining methods underscore the conservation of natural resources, as well as the comprehensive reclamation of the mining sites once operation ceases.

CORPORATE SOCIAL RESPONSIBILITY[4]

Unlike consumer products companies, in which competitive edge is obtained through two main processes: expanding the market for its particular product and ensuring a market environment conducive for growth, and successful corporate social impact through value chain development; in the case of ITM—being a mining industry—to sustain community acceptance and goodwill and consequent sustainable growth and profit, the company should ensure an environment conducive for sustainable operations and growth, and develop sustainable community resilience (employment, wealth, and health) in the area surrounding the mining operations during its operation and beyond, to ensure corporate image sustainability and risk mitigation.

ITM has long recognized that a company will only be successful if the surrounding communities are also developed alongside the growth of the company. So ITM has been focusing to enrich and develop the quality of life of the local communities since 2002. The "motto" of the company as regards to the interaction with the surrounding community is "Go Together, Grow Together in achieving Life Sustainability."

Following the Banpu Spirit, the strategy of the community development of ITM is as follows:

Innovation

The aim in this area is to improve quality of life through capacity building and development, and to establish community independence, particularly in education, health, and employment. This in turn will ensure sustainable community goodwill and a conducive operational environment.

Integrity

Community development is in line with the business strategy of ITM, and in compliance with the country's regulations and laws. ITM will continue its commitment and responsibility toward community development during mining operations and beyond.

Care

Local community empowerment to improve the standard of living, and to achieve a self-sustaining society, as well as the rehabilitation of the ecosystem are the main focuses of ITM care program. The future of the children living in the surrounding mining area will depend on the environment and their education, and ITM is committed to support them with a healthy living environment and basic education.

Synergy

Harmonious cooperation between the community, the government, and the company in implementing capacity building and education, and in sharing know-how within the context of community development, will ensure a sustainable result for life.

ITM has implemented development programs and community empowerment in the villages surrounding the mining areas. The pattern of program development and community empowerment is based on community involvement and participation through the formation of a community consultative forum (CCF) or community consultative committees (CCCs).

The sections to be covered in the community empowerment programs are as follows:

Economic Development

ITM activity encourages and facilitates communities in the development of local business, microfinancial institutions, entrepreneurship training, and improvement of infrastructure that support economic activities.

Social Development

Involvement in planning, implementation, and supervision in the following areas, including the necessary supportive infrastructures:

- Community education, by providing scholarships, improves local learning facilities and supporting educational programs. This includes training in good interactive teaching methods for primary schools and vocational studies such as computer courses, agriculture and fishery, simple accounting systems, quality assurance, and baking and cooking in cooperation with expert institutions. One of the popular programs is the training for motorcycle mechanics, conducted in cooperation with other parties (see the testimonial "Helping People Pursue Their Dreams").
- In the health sector, ITM provides health and medical facilities and is actively involved in improving infant nutrition.
- ITM actively promotes local arts and preserves local culture.

Environmental Preservation

ITM works to improve local waterways, land reforestation, and rehabilitation upon the cessation of its mine operations.

Community Relations

Through its CCF, ITM is closely involved in aiding and participating in the improvement of local and government facilities and other activities to cover the needs of the community in each mining site, which includes the celebration of local religious festivities.

Institutional Cooperation

To ensure optimum results of its community development programs, ITM cooperates closely with other institutions, such as the Corporate Forum for Community Development (CFCD), Indonesia Business Link (IBL), Indonesia Corporate for Sustainable Development (ICSD), Corporate Social Responsibility Indonesia (CSR+), Social Laboratory of University of Indonesia (LabSos UI), Social Laboratory of the Bogor Agricultural Institute (LabSos IPB), One Tambon One Product (OTOP) Thailand, and others.

Helping People Pursue Their Dreams

Asriadi is a young father of four small children living in the small village of Teluk Pandan, east of the mining site of PT Indominco Mandiri, one of ITM's subsidiaries. With very limited education and no resources, he spends his days patching tires, earning just enough cash to support his family.

In August 2008, Asriadi and 30 other people from selected villages surrounding the Indominco mining area attended an intensive motorcycle repair training, designed to provide basic knowledge and skills to amateur mechanics. The need for professional repair and service for two-wheeler vehicles had significantly increased, with the swelling number of motorcycles coming into their villages.

This training was a collaborative effort of Indominco as part of its CSR program, PAMA as a leading mining contractor in Indonesia, Kitadin, and Yayasan Dharma Bakti Astra (YDBA). As a followup program, Indominco and PAMA have also prepared a workshop model, managed in cooperation with the local youth clubs, offering technical and management training for mechanics who wished to set up their repair garage. With this arrangement, the association also support these new small entrepreneurs-to-be with the required funds.

Asriadi has been dreaming all along of owning a small repair workshop, and now with sufficient training and little working capital, Asriadi has come so close to this dream.

A. COMMUNITY PARTICIPATION AND EMPOWERMENT THROUGH THE CCF

Starting from the planning stages of the mining operations, ITM works closely with the local communities. This relationship continues up to the time of closure of the mines. ITM also works closely with the community, getting input and feedback directly from the local community in defining the right community development program to be funded and developed. These ideas are agreed upon and synthesized into the community action plans (CAPs). Through the company's CCC, ITM works with representatives from the government, community and social organizations to plan, implement, and supervise community development programs.

The locations and the CSR programs of CCC depend on the consensus of all the members of the committee, and are based on the needs of the community. The committee members are elected from the village community, while the head of the mining operation in each village, together with the highest local government head, act as co-advisors of the committee. All members are working as volunteers, and, as a team, work together in developing the village. One term for a CCC member is two years, after which period the committee can be re-elected.

Figure CS5.2 shows that, through CCC, social capital is expected to be generated through the involvement of three primary parties, namely the community, the company, and the government, which in turn will create and establish income sustainability for life. A concept of independence is induced, which means that the surrounding villages

FIGURE CS5.2 The pattern of community empowerment through CCC

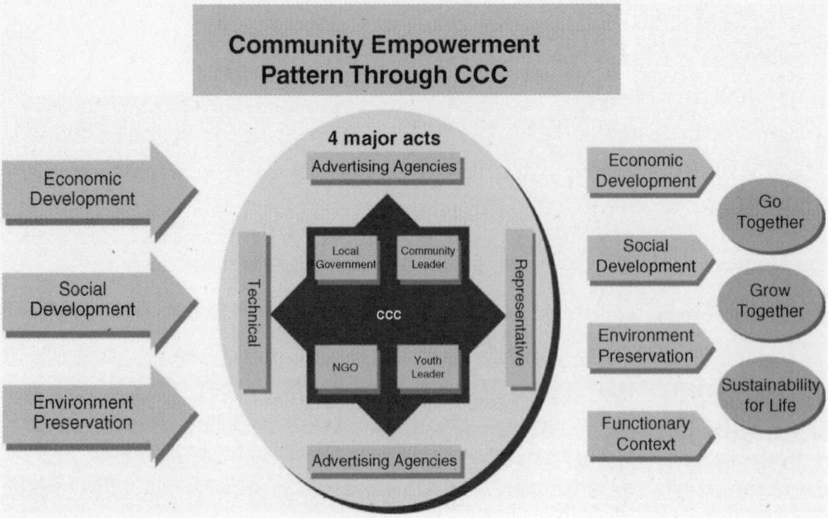

Source: ITM

generate and move social capital on their own initiative. Whenever a problem arises, surrounding villagers can independently align the initiative and keep its sustainability by using the existing social capital.

Case I—Leveraging Best Practices for Local Economic Sustainability

In sharing best practices in community development, ITM is able successfully to develop sustainable economy in sectors such as the home industry, fishery, agriculture, society, and environment. Particularly in developing the economic and social empowerment, ITM has defined three steps necessary to ensure the achievement of sustainable results.

Step 1: Provide the main infrastructure which cannot be provided by the locals for the people in remote areas.

Step 2: Encourage the development of social capital in the community through partnerships between the stakeholders based on available potential natural resources and local human resources.

Step 3: Develop economies of scale through local community aspirations, and awareness of the opportunities that increased resources available within the community will be supported by the local government.

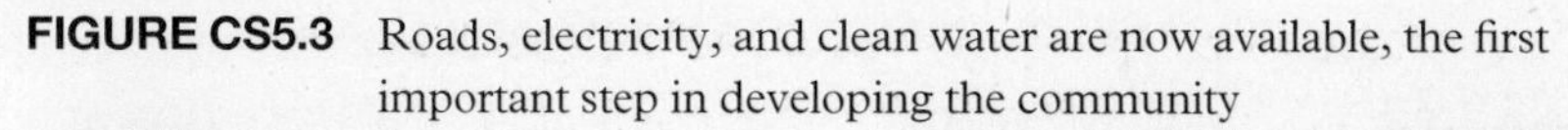

FIGURE CS5.3 Roads, electricity, and clean water are now available, the first important step in developing the community

Infrastructure Development

Infrastructure development is the basis for successful community development programs, which include the building of roads and bridges for easy public access, electricity supply in cooperation with the government (State Electricity Company), schools and vocational training for education and capacity building, medical centers and houses of worships, such as mosques and churches, equipments needed for each community development program, including home industries (see figure CS5.3).

Economic Development

The ITM mapping of economic development covers three main areas: agriculture, fishery, and home industry.

Trained Farmers Enjoy Doubled Crop Yields PT Indominco Mandiri (PT IMM), a subsidiary of ITM, operates its mining concession in a section with several villages within its Ring 1 zone, an area closest to the main mining site. Families in several of these villages earn their living mostly from traditional dry-land farming, cultivating vegetables, rice, and a few other crops. To increase the economic potential, ITM has created a special community program to develop the skills of these farmers, provided them with training to improve production, and also offered grants by way of agricultural machineries and infrastructure to farmers' groups.

Agus Salim is one of those farmers whose small field now generates better crop yields and increased income for him and his family. Together with his group, Agus Salim grows cauliflower, tomatoes, long

FIGURE CS5.4 A new hand tractor to help the farmers' group improve crop yields

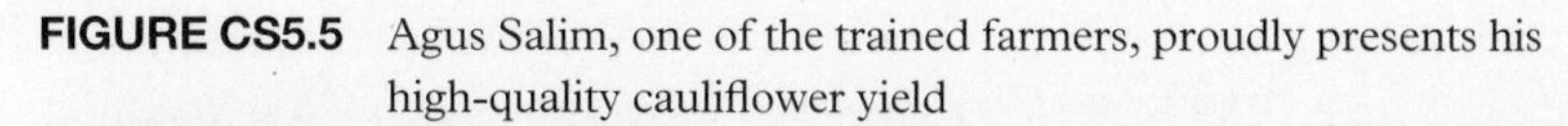

FIGURE CS5.5 Agus Salim, one of the trained farmers, proudly presents his high-quality cauliflower yield

© ITM

beans, squash, bitter melon, and cucumber. Traditional cultivation provided an average cauliflower yield at the time of harvest of about 110–220 pounds, but now that Agus Salim is equipped with new skills, his field can produce better-quality cauliflowers of up to 660 pounds. Recently, his group received a new hand tractor from PT IMM, and Agus Salim and his other farmer friends can expect to reap an even better harvest (see figures CS5.4 and CS5.5).

Seaweed A seaweed cultivation program was developed and launched in 2007 by PT Indominco Mandiri, a subsidiary of ITM. Seaweed cultivation potential in the island of Tihi-Tihi—an exotic small island a few miles east of the mining dock—is large. The good water ecosystem is ideal for growing seaweed, which will ensure the sustainability of this project. The company provides high-quality seeds, capital business training, and the necessary equipment. Currently, seaweed farming is developing fast, and is becoming one of the main source of

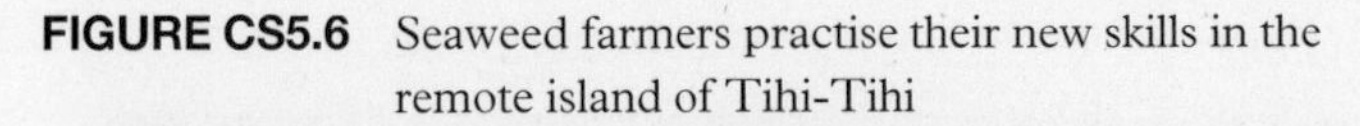

FIGURE CS5.6 Seaweed farmers practise their new skills in the remote island of Tihi-Tihi

© ITM

income for the island's residents. The income of seaweed farmers has increased dramatically from IDR400,000 to IDR500,000, (about $40–50) per one-and-a-half months.

Fishery Kitadin-Embalut is a program that has been initiated by a local breeder using holes created by PT Kitadin mining, which has been naturally filled with freshwater. After obtaining permission from the local government, and with the support from the Ministry of Agriculture and Fishery, the breeders use disbursement floats to cultivate fish for three or four months.

This program is growing fast and is creating significant opportunities for sustainable wealth creation. PT Kitadin has developed an integrated fishery development program which consists of a group of people cultivating fishery, and other groups focus on marketing and provision of seed and feed. In addition, PT Kitadin also encourages the establishment of home industry using fish as its raw material to produce snacks, fish paste, and so on.

FIGURE CS5.7 Fish cultivation is now a popular occupation

Rubber Plantation Rubber plantations in the areas surrounding the mining sites are cultivated and developed by another subsidiary of ITM, PT Trubaindo Coal Mining (see figure CS5.8). The company provides 53,000 rubber seedlings and pesticides. About 500 acres of land will be planted with the new rubber seedlings. The rubber plantation is a resilient community program, because the villagers are familiar in planting rubber using traditional equipment, and rubber is also a good replacement of the traditional rattan plants from Kalimantan, the price of which has decreased by 30 percent.

Home Industry (Vocational and Educational Training) A program called Home Industry Team Indominco Mandiri supports vocational studies and capacity building for unproductive local residents.

With the consideration of the local environmental biodiversity, women and housewives are trained to become entrepreneurs. They

FIGURE CS5.8 The company facilitates community-run rubber cultivation by providing seedlings, pesticides, and training

are facilitated to produce and market all kinds of food, snacks, and beverages from local agricultural and fishery products. All these products receive certification from the government and are marketed and distributed in the nearby city of Bontang and its surrounding villages.

The program supports the young people through the establishment of ADEGA (Academy of Design and Graphics) in which they can learn how to do "digital printing" for banners, flags, T-shirts, and so on.

It also encourages the locals to develop their skills to become good tailors and dressmakers. It has also established the APIM (Indominco Tailor Association).

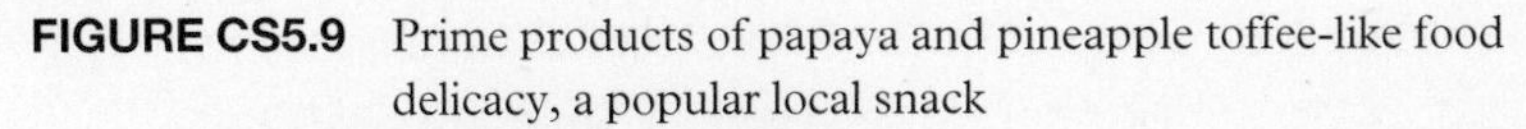

FIGURE CS5.9 Prime products of papaya and pineapple toffee-like food delicacy, a popular local snack

© ITM

All these CSR activities are categorized as self-employment and wealth creation, which will ensure sustainable income for the community, even after the closure of the mine operations, and consequently community resilience and independence for the future.

Case II—Integrated Development of Tihi-Tihi Island

Tihi-Tihi Island is a small fishing island, located about one hour's traveling time by boat from the mining dock, Tanjung Laut Habour, Bontang, Kalimantan. Tihi-Tihi is a remote rural area with no clean water. To get clean water, the community must spend IDR3,000 (about 30 cents), which for them is pricy, to get 30 pints of clean water for drinking. To bathe and wash, the people just use seawater. No schools and no medical centers are available on the island.

PT Indominco Mandiri, which is a subsidiary of PT ITM, has recognized the island's potential natural resources. The availability of abundant fish in the waters surrounding the island and a good water ecosystem for farming seaweed could become a source of income and could create employment and wealth for the local community.

At the same time, it has been recognized that there is a need to have school and training facilities for education and capacity building of the community. In line with the CSR policy of the holding company, PT Indominco Mandiri has implemented the four CSR categories. The CSR activities implemented were improvement of quality of life of the local community by:

- ensuring enough supply of clean water. The company donated a motor boat to transport water from Tanjung Laut to Tihi-Tihi, and also built five water reservoirs with a total capacity of 7,400 pints, enough clean water for 52 families for two days
- improving schooling facilities for elementary schools
- ensuring the availability of an integrated community health center
- funding programs that provide working capital, and also provide access to the wider market for the farmers to sell their products
- capacity building through education by ensuring the availability of competent teachers, offering scholarships, and improving schooling facilities
- rebuilding the availability of social capital to support wealth creation for the community
- involving the community, through CCC, in defining their future and the development of their community and the area in which they live
- ensuring good cooperation among all stakeholders to get quick, harmonious, and accurate results from the community development initiatives (see figures CS5.10, CS5.11, CS5.12).

FIGURE CS5.10 An isolated island now closer to the world

FIGURE CS5.11 A girl walking to her newly built school in Tihi-Tihi Island

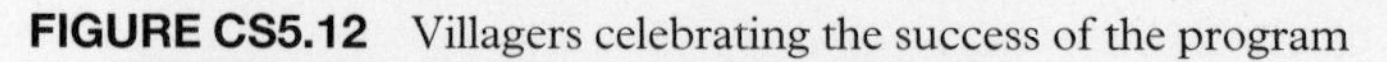

FIGURE CS5.12 Villagers celebrating the success of the program

© ITM

B. HIGH-PERFORMANCE COMMUNITY DEVELOPMENT PROGRAM

With five mining sites operating in three different regions, and each having its own challenges, it is essential to increase the competencies of the community development officers (CDOs) and the CCC leaders and members. ITM organizes a half-yearly meeting with the CDOs and meets with the CCC leaders and members in a formal meeting called CCF held in each mining location. Issues ranging from changes in community aspirations, stakeholders characteristics, and CSR program achievements, as well as guidelines in handling structural and psychological challenges are discussed.

To improve communication and to monitor the efficacy of CSR programs in those various locations, ITM has developed an integrated monitoring system called *Community Development Management Information System* (CD-MIS), gathering and sending information from sites to the head office (see figure CS5.13). Data input covering

FIGURE CS5.13 An overview of ITM's CD-MIS

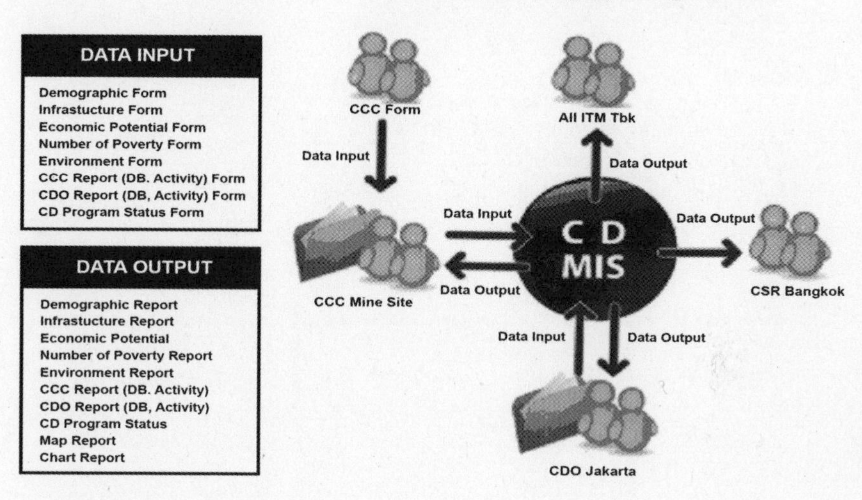

Source: ITM

demographic information and basic infrastructure data, economic potential, poverty figures, environmental records, CCC and CDO Reports, and the community development program status makes up the basic of the system. The output is sent to various offices' locations, including the head office in Bangkok.

C. PLAN OF CLOSURE OF MINING OPERATIONS

In accordance with the Ministry Decree on the closure of the mining operations, ITM and all its subsidiaries have developed the Mining Closure Planning Documentation for submission in May 2009. Technical and nontechnical, including the social and economic groundwork and research are done with the support of consultants from major national universities. One of the main efforts focuses on developing the criteria for a successful community development program. This covers all stages, from planning, implementation, evaluation, recommendation, to monitoring.

To ensure successful and sustainable mine closure implementation, all sites consulted the local governments for effective coordination. The

CCC in each location is actively involved in gathering local data, taking into account community resources, social dynamics, local culture, and government programs applied in the area.

NOTES

1. PT Indo Tambang Megah, Tbk (ITM), Annual Report 2008.
2. See successful case studies of ITM.
3. http://www.banpu.co.th/eng/governance/cg_2008.php.
4. http://www.banpu.co.th/en/07-social-responsibility/ index.php

Case Study 6

H.J. HEINZ[1]

THE COMPANY

Heinz is a global food company built upon a legacy of integrity, ethics, transparency, and community engagement. From Pittsburgh, Pennsylvania, where the company is based, Heinz markets more than 16,000 products globally, primarily in three core categories: Ketchup and Sauces, Meals and Snacks, and Infant Nutrition. In fiscal year 2007, Meals and Snacks accounted for more than 44 percent of the company's total sales, followed by 41 percent in Ketchup and Sauces, and the remaining 15 percent in Infant Nutrition.

Reflecting Heinz global presence, more than half of the sales are outside North America. Its fiscal 2007 report stated annual sales of $3.1 billion in Europe, $1.2 billion in Asia–Pacific, and $427 million in other regions outside North America. To drive growth, Heinz has been investing in emerging markets, including Russia, India, China, Indonesia, Poland, and Latin America. These emerging markets are growing at double-digit rates and generating almost one-third of the company's sales growth. The expectation is that the total sales of the emerging markets will be approximately $1 billion in fiscal 2008.

CORPORATE GOVERNANCE AND ETHICAL OPERATION

Heinz has established a firm system of corporate governance to ensure the company's accountability to shareholders and the community; and it operates in an ethical and socially responsible manner.

The Heinz Global Operating Principles express the company's unwavering commitment to a safe and fair treatment of all employees, to environmental protection and to the respect of the cultural, ethnic, religious, political, and philosophical differences among people.

CORPORATE SOCIAL RESPONSIBILITY

As a global food purveyor, Heinz recognizes that everything it does to bring consumers nourishment and nutrients, from the seed to the store, strongly affects people, communities, and the environment. Heinz has a firm commitment to corporate social responsibility (CSR), and has embedded this principle in the way they conduct business. This is clearly reflected in Heinz Corporate Social Responsibility Mission below:

> Heinz will achieve sustainable growth by enhancing the nutrition, health, and wellness of people and their communities. We will conduct business in an ethical manner, guided by our strong commitment to integrity, safety, and the principles of social and environment responsibility, in communities where we operate. Heinz will manufacture safe and high quality products, reduce environmental impacts, and maintain business and labor practices that ensure human safety and dignity. As a global company, Heinz will also make the world a better place to live by helping people in need through our charitable foundations, employee contributions, and community partnerships.

A. CORPORATE IMPACT TO SOCIETY THROUGH THE DEVELOPMENT OF THE VALUE CHAIN

Case I—Premier Hybrid Processing Tomato Seeds

At Heinz, it starts with tomato seeds, but ends with products that make Heinz the world's leading producer of processed tomato products. Heinz's expertise in tomatoes dates back to the origin of the company, when founder Henry J. Heinz introduced his brand of tomato ketchup in 1876.

To ensure significant competitive advantage, Heinz has done the following:

- developed successfully premier hybrid processing tomato seeds, delivering the best-tasting, highest field performance, and the best consistency of tomato varieties in the world. Heinz also developed hybrid tomato varieties that can adapt to various global climates to benefit growers, processors and end-users everywhere. Heinz is not only selling these hybrid tomato seeds to affiliated growers, but also globally through affiliates and international dealers. Heinz-Seed is the market share leader in North America and the world.
- established processes to ensure that it can trace the company's tomato supply from the seed to the store. Heinz is able to track where the tomato seed originates, what is used to help grow and fertilize the tomato plant, where the plant is grown and how, when, and where the tomato is processed into finished product and how it reaches its final retail distribution.

Heinz has worked closely with the California Tomato Growers' Association and the Processing Tomato Foundation, in conjunction with the University of California Davis to develop guidelines for sustainable tomato production and processing in California. Also, being a member of the Business Coalition of Sustainable Food Lab., whose key objective is to promote sustainability throughout the supply chain, and to maintain balance among people, profits, and environment, this coalition promotes creative solutions to supply chain concerns and discusses issues in a constructive way to find solutions.

Heinz has used its knowledge to benefit the world, that is, China, Southeast Asia, India, Europe, the Middle East, Africa, Caribbean nations, and Latin America.

Currently, with 8,000 domestic and 35,000 international suppliers, Heinz has ensured that its suppliers, co-packers and joint venture partners uphold uniformly high standards of quality and services by introducing the Heinz's Supplier Guiding Principles. In 2007, Heinz has established a global supply chain task force. The primary objective of the task force is to increase the company's competitive advantage by improving coordination and reporting systems across the supply chain and to harmonize global processes and benefit from Heinz's scale among suppliers.

B. DEVELOPING AND EXPANDING THE MARKET THROUGH COMMUNITY DEVELOPMENT AND IMPROVING QUALITY OF LIFE AND CHANGE OF HABIT

Case II—Heinz Micronutrient Campaign

Heinz donated more than $17 million in cash and products across the globe over the past two fiscal years to support community programs, with a focus on promoting the health, nutrition, and wellbeing of children and families. Heinz believes that all children, no matter where they live, deserve a fighting chance to grow up strong and healthy. To support the introduction of its global infant nutrition products, Heinz, together with its nonprofit foundation, is committed to tackling micronutrient malnutrition through the Heinz Micronutrient Campaign.

According to the World Health Organization, the three most common forms of micronutrient malnutrition, which are iron, vitamin A, and iodine deficiencies, cause more than 50 percent of child deaths worldwide each year. At the same time, it also perpetuates poverty and costs deprived countries of as much as 3 percent of their yearly gross domestic product. The good news is that malnutrition can be prevented. Heinz has supported the development of "Sprinkles"—an inexpensive, dry, tasteless, single-serving packet of iron, vitamins C, D, and A, and zinc mix,[2] which can reduce the global surge in malnutrition drastically. The effectiveness has been proven because children taking "Sprinkles" showed marked improvement in overall health. Since its introduction, more than 1.2 million children in Ghana, Guyana, Haiti, Indonesia, Mongolia, and elsewhere have benefited from the Heinz Micronutrient Campaign.

To ensure successful operations devoted to the health and wellness of consumers, particularly children, Heinz recognizes the need to cooperate with local governments, nongovernmental organizations (NGOs), and other international and country's health organizations. In Indonesia, Heinz developed a new infant nutrition product formulation called *Vitalita*, adapted to the local staple foods, under a program named Project Mayang in 2003. Heinz and its foundation, in cooperation with Helen Keller International, an international NGO, trained public health workers on the benefits of the Heinz Micronutrient Campaign, and had them administer the distribution of 44 million sachets of *Vitalita* to the community. Hellen Keller International also evaluated and monitored the effectiveness of the program. Similar operations have been done in China (Project Mei Li, 2007) and in India (Project Ashok, 2007).

C. ENVIRONMENTAL CARE PROGRAM

From sustainable agriculture and energy-efficient manufacturing processes to eco-friendly packaging initiatives, Heinz is proving that smart business is compatible with environment stewardship.

Heinz is focusing its environmental programs in four key areas: waste, water, energy, and greenhouse gas emissions. Significant results in waste reduction are achieved through packaging innovation and recycling. Customer and supplier collaborations are among the driving forces affecting waste reduction and packaging optimization initiatives, for example, all cans from products such as beans and soup have been converted to lighter-weight cans (the UK, Australia, Indonesia, and New Zealand).

D. PHILANTHROPY

The Heinz Company Foundation was established in 1951, committed to promote health, nutrition, and wellbeing of children and families, and also education, youth services, diversity, and quality of life globally.

Heinz has an enduring legacy of providing humanitarian aid to communities affected by natural disasters, for example, calamity caused by Hurricane Katrina, earthquake in Yogyakarta, or big floods in Jakarta.

E. EMPLOYEES' VOLUNTEERISM

Heinz employees are encouraged to support charitable organizations actively through many programs and widespread campaigns that the company is supporting, including the Heinz Micronutrient Campaign, United Way, and special events such as the Heart Walk.

The foundation oversees a volunteer group called Heinz HELPS, which stands for Heinz Lending Public Service. Heinz HELPS facilitates employee involvement and participation through a grassroots approach.

The program uses an internal website that enables employees to collaborate on projects and coordinate volunteer walks and programs.

Recognition for CSR

Heinz's strong performance across a broad range of social and environmental criteria is widely recognized, as can be seen in the company's 2007 Corporate Social Responsibility Report:

> Heinz is a public company whose common stock (NYSE:HNZ) trades on the New York Stock Exchange. We are proud of our inclusion in three high-profile indexes—the Dow Jones Sustainability Indexes, the Calvert Social Index and the Domini 400 Social Index—that track the financial performance of companies that demonstrate excellence in their attention to and management of environmental, social, and governance issues. In 2007, Heinz received the highest score from investment analysts in The LOHAS Index™ ranking of 50 most environmentally and socially responsible companies. Heinz was also ranked among the 100 Best Corporate Citizens for 2007, a list published by CRO (Corporate Responsibility Officer) Magazine. Approximately 1,100 publicly-held U.S. companies in the Russell 1000, S&P 500, and Domini 400 indexes were considered with rankings based on extensive data collected by KLD Research and Analytics, an independent investment research firm. This recognition reflects Heinz's strong performance across a broad range of social and environmental criteria and our return to shareholders.

NOTES

1. H. J. Heinz Company Corporate Social Responsibility Report, 2007.
2. Created by Dr. Stanley Zlotkin of Toronto's Hospital for Sick Children.

Case Study 7

TNT INTERNATIONAL B.V., THE NETHERLANDS

THE COMPANY[1]

TNT is a global mail and express company, which operates in 200 countries and in many different types of environments. The company was established in 1946, and built its reputation on a record of commitment to meeting customers' needs. From only one truck, when Ken Thomas started his own transport business in Australia, now TNT Express employs more than 75,000 people and operates 26,000 road vehicles and 47 jet freighter aircraft, and a network of more than 2,300 company-owned depots.

As a responsible employer, TNT ensures that operations are done safely. TNT has always focused on embedding a strong safety culture, top down and bottom up. TNT recognizes that managing health and safety risks cannot be treated in isolation from other company processes. Therefore, these factors are integrated into all TNT activities. Health and safety concern is the primary responsibility of all TNT managers, from senior executives to the first-line supervisors throughout the company. Specific health and safety aspects are delegated company-wide using the cascade principle and are recorded in job descriptions. It is TNT's policy to ensure that all operations meet the corporate sustainability standards.

As a leader in global express service, TNT feels the responsibility to be actively involved in the reduction of CO_2 emissions. This is one of TNT's main concerns. TNT has acknowledged that inland and air transport accounted for 18 percent of global emissions, and its line of business unquestionably contributes to the production of CO_2 emissions. In 2007, TNT was accountable for the emissions of 1,000 kilotons of CO_2. If the figure was to include TNT subcontractors, the number would increase to 2,500 kilotons. It was in the same year that TNT launched a comprehensive program to cut CO_2 emissions.

CORPORATE SOCIAL RESPONSIBILITY

To support the company mission, TNT seriously embraces environmental and social obligations. The company seeks to lead the industry by instilling pride in their people, creating value for stakeholders and sharing responsibility for the world. Corporate social responsibility (CSR) fits into these values. TNT is in the lead in the reduction of CO_2 emissions efforts, and has also established partnerships with a range of charitable and relief organizations.

TNT, in 2007, topped the Dow Jones Sustainability Index (DJSI), a prestigious index that tracks the financial performance of leading sustainability-driven companies. DJSI covers the top 10 percent of the biggest companies under the Dow Jones Index in terms of economic, environmental, and social criteria.

A. CORPORATE PHILANTHROPY AND EMPLOYEE VOLUNTEERISM

Case I—Fighting World Hunger

In December 2002, the United Nations World Food Program (UNWFP) and TNT launched a five-year partnership aimed at a single common goal: fighting world hunger. UNWFP is the world's largest humanitarian agency and is the food assistance arm of the UN. Those five years have proven to be a success for both organizations that they renewed their commitment in 2008.

UNWFP logistics is at the core of its operations. It deals with logistics needs during humanitarian crises for both food and nonfood items, with an aim to improving and increasing the overall industry response and ultimately to save lives. Since the start of the partnership, TNT has contributed a total of €32.5 million in-kind to UNWFP operations and €5.5 million in cash. Besides this contribution, €9.0 million was raised by TNT employees and donated to WFP. This donation went toward supporting the UNWFP School Feeding Program in Cambodia, the Gambia, Malawi, Nicaragua, and Tanzania. In these countries, 27 TNT employees volunteered to spend three months working side by side with UNWFP on school feeding projects, where they witnessed the impact of the donations on children's lives first hand.

B. GAINING COMMUNITY ACCEPTANCE AND GOODWILL THROUGH CARE FOR THE COMMUNITY AND THE ENVIRONMENT

Case II—Planet Me: Becoming a Zero-Emissions Transporter

The response of TNT toward emission reduction is by initiating a new program called Planet Me in 2007, with a vision of becoming a zero-emissions transport company. Planet Me aims to ensure that TNT manages its contribution to the effect of global warming effectively. TNT has divided this project into three separate components: Code Orange, Choose Orange, and Count Carbon.

In addition to the obvious reasons for reducing carbon emissions, TNT is convinced that taking care of the environment is crucial for business. Surveys show that customers share the same environmental concerns. In fact, they evaluate TNT and its competitors increasingly on what TNT does about climate change.

Code Orange

This is a mandatory program that covers every aspect of TNT business, putting environmental stewardship at the heart of the company's strategy and culture. Code Orange looks at the actions that TNT can take as an organization to reduce CO_2 emissions. It focuses on eight key initiatives: Aviation, Buildings, Business Travel, Company Cars, Green Investments, Operational Vehicles, Procurement, and Partnering with Customers.

Choose Orange

The estimated CO_2 footprint of more than 160,000 TNT employees and their families is larger than those of the entire company and subcontractors combined. As a voluntary program, Choose Orange extends the Planet Me commitment to those areas of company's influence: the homes of its employees. TNT has started to look for ways to reduce personal carbon footprints at home, which often saves money at the same time. There are a lot of inspiring initiatives from TNT employees worldwide, and these are often implemented by themselves based on their own ideas. Examples are aplenty; TNT Portugal, with its home-visit electricity and fuel consumption analysis and reduction program called *Quercus*; TNT Bulgaria with its Ride A Bike, Be On Time program, which rewarded employees who leave their cars at home and ride free orange bicycles instead; and TNT personnel association in the Netherlands who is negotiating for a major discount on green energy for all 80,000 employees, which can reduce the global TNT family footprint significantly.

Count Carbon

Count Carbon is about managing, monitoring, and reporting on CO_2. This project within Planet Me aims to improve the understanding of emission sources and the potential of measures. Its main objectives seek to determine the magnitude of subcontractor emissions and to project emissions into the future to generate a complete view of TNT's carbon footprint and the related challenges. This understanding is important to identifying the areas where TNT can make the most significant changes.

TNT has also identified its potential for reduction. In 2008, the biggest win lies in increasing fuel and energy efficiency and expanding green electricity sourcing. For 2009 and beyond, TNT subcontractors offer the biggest reduction potential, at the same time, the largest challenge. With Count Carbon, TNT focuses on further improving its CO_2 management, monitoring, and reporting practices. This will allow the company to track and manage emission systematically and improve CO_2 performance.

NOTE

1. http://www.tnt.com/express/en_id/site/.

Case Study 8

INTEL CORPORATION, US

THE COMPANY[1]

Intel Corporation is the world's largest semiconductor manufacturer, based in Santa Clara, California, the US. The company was founded in 1968 by two semiconductor pioneers, Gordon Moore and Robert Noyce, with the main intention to pursue large-scale integrated (LSI) memory. Andy Grove soon joined this new venture, and his strong leadership skills made him the company's key business and strategic leader.

Intel has led the industry in the development of many significant innovations as early as in the year the company was born, including the first microprocessor in 1971, the invention that sparked a revolution. With the passion to create technology that change the world, Intel attracts the most brilliant minds in science to push the boundaries of innovation and further its position as the leader in semiconductor technology.

Its 2008 annual result reported that Intel posted revenue of $37.6 billion, with operating income of $9 billion, net income of $5.3 billion, and earnings per share of 92 cents. Intel generated approximately $11 billion in cash from operations, paid cash dividends of $3.1 billion, and used $7.1 billion to repurchase 324 million shares of common stock. With the uncertainty in global economic conditions, the company's president and CEO, Paul Otellini, said that Intel will continue to extend its manufacturing leadership, drive product innovation, develop new markets, and implement operating efficiencies. The company's fundamental business strategies are more focused than ever.

CORPORATE SOCIAL RESPONSIBILITY

Intel believes that corporate responsibility is an essential part of the business. The success of the company cannot be separated from the health of the planet and the communities where it operates. As a global technology and business leader, Intel considers its stakeholders—neighbors, suppliers, customers, policy makers, employees, analysts, stockholders, and others—to define the major corporate responsibility focus areas that cover economic sustainability, environmental stewardship, and improvement of education and technology access across the world. It has been the company's success in the past, and it will continue to be in the future.

At Intel, it is fundamental to conduct everything in the right way. This includes the setting of high ethical expectations for its employees and suppliers, and the provision of a progressive and inclusive workplace to deliver products that can change the world. Intel has focused for a very long time on reducing emissions, recycling waste, conserving water, and designing products with the environment in mind. Intel has also launched community programs that combined technology, volunteerism, and financial support to transform lives worldwide.

A. GAINING COMPETITIVE EDGE THROUGH EDUCATION AND LEVERAGING KNOW-HOW

For Intel, the strategic benefit of increasing community education is very critical. The success of the company depends on skilled engineers and innovators, a healthy technology ecosystem, knowledgeable customers, and thriving communities. Intel believes that students deserve the skills they need to be part of the next generation of innovators.

Giving a child hands-on access to computers and the internet can change the course of his or her life. Equipping a single teacher with the skills and resources to increase the effective use of technology in the classroom can affect hundreds of students. Providing university faculty members with a cutting-edge curriculum can alter the impact that their students will have on technological advancements for years to come.

For these reasons, Intel has formed partnerships with governments, ministries of education, universities, and nongovernmental organizations with the goal of enabling education. Over the past decade, Intel has invested more than $1 billion and its employees have donated more than two million hours toward improving education worldwide. Through its education programs, Intel provides opportunities for future innovators.

Case I—Intel Teach Program

This program helps teachers around the world integrate and create active learning environments in the classroom. Through Intel Teach, the company provided professional development for more than 1.1 million teachers in 2007, bringing the total number of teachers trained to more than 5 million since the program's inception. Intel Teach is now available in more than 40 countries, including six new countries in 2007: Indonesia, Libya, Peru, Romania, Sri Lanka, and Trinidad and Tobago.

Case II—Intel Learn Program

Intel Learn is an informal after-school program that helps students in developing countries build technology skills through hands-on projects. The program, offered in government-funded community technology centers, is currently available in nine countries: Brazil, China, Chile, Egypt, India, Israel, Mexico, Russia, and Turkey. An independent evaluation from a nonprofit research firm SRI International has showed that learners who complete the program demonstrate improvement in technology literacy, collaboration, and critical thinking skills.

Case III—Intel Computer Clubhouse Network

The Intel Computer Clubhouse Network is a community-based after-school program that brings technology to youth in underserved areas in more than 100 locations in 20 countries. Computer clubhouses offer an environment of trust and respect where young people can develop technology fluency, collaborative work skills, and a sense of their own potential.

Intel's other education initiatives are the Intel International Science and Engineering Fair and Intel Science Talent Search; both are programs of the Society for Science and the Public. Through these programs, high-school students complete original research projects and compete for millions of dollars in scholarships and awards each year.

NOTE

1. http://www.intel.com.

Case Study 9

MOTOROLA

THE COMPANY[1]

Motorola is known as a leader in global communications, and has been in the business of communication inventions and innovations for the past 80 years. In 2007, Motorola acquired Symbol Technologies and realigned its business to cover the areas of technology, solutions, and services within its three primary segments of Enterprise Mobile Solutions, Home and Networks Mobility, and Mobile Devices.

The beginning of Motorola's expertise in mobile communication was the car radio, manufactured by Galvin Manufacturing Corporation in 1930, which was soon adopted for wider use by the police departments and city governments across the US for public safety use. Later, after the introduction of the two-way radio system Police Cruiser Radio Receiver, Galvin Manufacturing Corporation developed Handie-Talkie Radio, a handheld communication device during World War II. After its first public stock offering in 1943, the company changed its name to Motorola, Inc. in 1947.

Motorola, as a pioneer in mobile technology, is proud of the company's heritage, which is rich in technological breakthroughs and innovations. The company also played an important role in the Apollo Space Program, where a Motorola transponder transmitted communications between Earth and the Moon. The world's first commercial portable cellular phone, the Motorola DynaTAC phone, revolutionized how people perceive mobile communications.

The company now operates in 72 countries, with net sales of $36,622 million in 2007. The US is the largest market for Motorola products, while mobile devices, that is, mobile phones and accessory products, dominate its worldwide sales.

CORPORATE SOCIAL RESPONSIBILITY[2]

In a September 2008 press release, Motorola announced its social recognition, ranked as Sustainability Leader by Dow Jones Sustainability Index, for the company's environmental, social, and economic performance for a fifth straight year. This is proof of Motorola's commitment to corporate responsibility, the commitment to conduct business with integrity.

Operations at Motorola are guided by a set of codes of business conduct that describes the ethical standards required of employees. This was developed based on the company's values of doing the right thing. The following principles are at the core of every action taken:

- Focus on innovative products, customer delight, and quality
- Operate with high standards of ethics and transparency
- Encourage environmental quality and sustainable use of Earth's resources
- Promote diversity and inclusion
- Maintain a safe and healthy workplace for all employees
- Create wealth, economic opportunities, and growth in regions where the company operates
- Maintain supplier relationship and suppliers' compliance to applicable laws
- Support the community
- Preserve shareholders' through sustained profitable growth, technological innovation, and market leadership

In line with the company's commitment to operate transparently, Motorola publishes its yearly Corporate Responsibility Report with the use of the Global Reporting Initiative (GRI) reporting guidelines, including the reporting of the company's social and environment-related activities. As a global company ringing up $36.6 billion in sales,

Motorola undeniably makes a significant impact on the global economy and societies around the world.

Motorola exercises this opportunity to bolster communities by sharing financial resources and products, and engaging its employees in supporting charitable and civic organizations with their time and money. The focus of Motorola's community investment is in supporting basic education, in providing access to communication technology, especially in the developing countries, and in providing relief for victims in disaster areas around the globe. Through the Motorola Foundation, funds are strategically granted to create strong community partnerships, fostering innovation and engaging stakeholders, which basically is the mission of the foundation.

Social and Environmental Impact on Community Through Supply Chain Development and Sharing Best Practices[3]

As a global technology company, Motorola manages a complex supply chain with more than 27,500 suppliers located in 126 countries.

Investing in developing countries as a strategy to move part of its manufacturing to these countries has many implications: economically, socially, and environmentally. The investment indeed contributes to the economic development, as well as provides new job and wealth opportunities. However, there are several issues concerning the working conditions, compliance with the law, and also those related to environmental protection in these countries. It is critical for Motorola to spread its corporate responsibility principles throughout its supply chain to ensure high quality and continuity of supply.

Motorola reported that the company is actively involved in dealing with supply chain issues at an industry level, and joining forces with other electronics companies to drive greater improvement to supply chain conditions. Business conduct expectations are incorporated into supplier contracts as a means to spread best practices, and Motorola provides continuous development through training, and monitoring and expanding suppliers' capabilities. Suppliers' code of conduct requires the suppliers to adhere to Motorola values and policies that include:

- compliance with the law
- rejection of corruption
- avoidance of discrimination
- nonuse of forced or child labor
- permitting workers to choose or join an association
- avoidance of excess overtime
- payment to workers of at least the legal minimum wage (or industry standard, where no minimum wage law exists)
- operation of a safe and healthy work environment
- operation of an environmental management system
- disclosure of materials in the products supplied.

Motorola engages the management, the industry, and its stakeholders in the development of reliable suppliers with the goal of establishing long-term business relationships.

Motorola has also established a monitoring system to improve the performance of its supply chain continuously, and also to identify and avoid suppliers with conflicting practices (see figure CS9.1).

Motorola wants its suppliers to succeed and helps them to correct issues identified through audits, providing feedback, and training to enable the suppliers to build their own policies and compliance processes.

Expanding the Market Through Community Investment[4]

Based on Motorola's Corporate Responsibility Report, in 2008, Motorola and the Motorola Foundation invested $23.7 million in community-based programs, with an addition of $4.3 million coming from employees' donations. The focus of Motorola's community investment is in three areas, namely education; projects that provide access to communication technology primarily for the developing countries; and disaster relief assistance. Being a technology company, Motorola supports basic education programs primarily in science, technology, engineering, and mathematics (STEM). These programs are supported by cash and product donations, as well as by employee volunteerism.

> "Is community investment a distraction from core business activity?"
>
> No. Our community investment benefits Motorola by strengthening local relationships, improving employee morale, and boosting our reputation. Our support for science, technology, engineering and math education helps address skills shortage in these areas and build skills we—and all of society—will need in the future.
>
> —*2008 Motorola Corporate Responsibility Summary Report*

In preparing the next generation of innovators, in 2007 alone, Motorola offered $3.5 million in a program called Innovation Generation Grants, a program for young people to learn more about STEM and to develop interest and skills in technology-related fields, especially for girls and underrepresented groups. In 2008, the plan was to increase the funding, and expand the coverage outside the US. In supporting education around the world, Motorola sponsored funding for science clubs in Brazil, Chile, China, India, Israel, Mexico, Russia, and the US. Since 1994, Motorola and the Motorola Foundation have helped more than 23,000 children in the rural areas of China to return to elementary school, and have donated more than $5 million for a project called Project Hope—a program developed by the China Youth Development Foundation—this includes funds for Motorola Hope Schools, teachers' training, resources, and students' scholarships for advanced education.

NOTES

1. http://www.motorola.com.
2. http://www.motorola.com/responsibility.
3. Motorola Corporate Responsibility Report 2007 and Motorola Corporate Responsibility Report 2008.
4. Ibid.

Case Study 10

CSR IN THAILAND[1]

THE COUNTRY

Thailand is an independent country, a constitutional monarchy that is different from its neighbors in Southeast Asia, in that it has never been colonized. The governmental system consists of the King of Thailand, the parliament, the Royal Thai Government, and the Office of the Prime Minister. Although the government is run by the Office of the Prime Minister, with the prime minister as the head of government, but the influence of the king, who serves as the head of state, is still very strong. Currently King Bhumibol Adulyadej is the longest-reigning king and is very much loved and respected by all his people in Thailand.

Thailand's economy is supported heavily by exports of agricultural and fishery products, goods, and services, which account for about two-thirds of the country's GDP.[2] Thailand is the world's biggest rice exporter and with a major portion of the population working in agriculture, the Royal family in collaboration with the government engages in many sustainable development initiatives that encourage food and economic sufficiency for small farmers as well as at the national level.[3]

CORPORATE SOCIAL RESPONSIBILITY IN THAILAND

The modern concept and methodologies of corporate social responsibility (CSR) have come to Thailand through multinational companies implementing various activities according to each corporate strategy.

However, in Thailand, with Buddhism as the major religion and the king being the official upholder of the religion, the principles of social responsibility are part of the traditional beliefs. Buddhism is strongly rooted in the heart of Thai people and organizations.

The king of Thailand has been known for his deep commitment and devotion to his people. During his long reign of 60 years, the king and the royal family have initiated and undertaken many social and economic development projects for most of their people who are poor, to improve their quality of life and help them become self-sufficient. The king has spent most of his years to travel and visit the provinces and rural areas of the country, to understand the needs and the problems faced by his people. The development activities are implemented through various projects known as royal projects that serve as a great model for companies aiming to apply CSR with sustainable social, environmental, and economic results for the community.

Royal Project Foundation

The idea of the project started in 1969. The king was concerned over a long tradition of opium poppy cultivation as the main source of income for hill tribe families in the highlands of northern Thailand. Although cultivation of opium poppy was at the time already banned, farmers had no better alternatives. The traditional agricultural method of slash and burn to clear the land for poppy farming also destroyed forests and watersheds, bringing floods to the plains, and threatened the environment.

The king of Thailand, His Majesty King Bhumibol Adulyadej, who is always respected for his attentive understanding, won the hearts of the hill tribe families who then agreed to grow substitute crops. This was followed by the king deciding on a program to bring worthy livelihood to these farmers, supported by an integrated assistance of much needed highland agricultural research, education, and capability building. This became the Royal Project with main focus on research and development during the first decade. From the beginning, the project has pursued its mission in collaboration with the Royal Thai government, universities, public and private agencies, international organizations, and volunteers.

The work of the Royal Project is based on its main objectives, which, besides eradicating opium poppy cultivation, include conserving the natural resources and also offering a helping hand to all humankind. The project originally developed to help the hill tribespeople was later known as the Royal Project Foundation, covering four major areas of research, agricultural extension, development, and socioeconomic activities. It has now an effective management system and permanent budget allocation.

Today researchers test hundreds of temperate-climate fruit trees, vegetables, flowers, tea, coffee, and herbs in 28 agricultural experimental stations in the hills to assess their potential as cash crops. Farmers are also introduced to agricultural technology and farming without destroying the environment, and they are able to produce high-quality fruits and vegetables for local and overseas consumers. The project also involves the local people in the production of processed and canned fruits like jams and wines, frozen strawberries, canned vegetables, and dried fruits and flowers for export. Nearly 300 upland villages benefit directly from the Royal Project, which is also introducing schools, cooperatives, rice banks, and primary medical services.

Currently, opium cultivation has declined by 85 percent, and the farmers have transformed into productive vegetable, fruit, coffee, and flower growers. Through international cooperation and high national commitment, the Royal Project is helping to reduce the supply of illegal drugs. The project has helped the hill tribe families to help themselves and enabled them to have a better standard of living and sustainable prosperity.

The Chaipattana Foundation

A site dedicated to His Majesty the King of Thailand for his 60 years of inspiring hard work quoted the king as a philosopher who "has devoted his life and resources to aiding the development of the Kingdom and the improvement of the Thai people's livelihood. He has initiated thousands of innovative agriculture, environment, health, and education programs to raise the standard of living in Thailand."[4]

One important organization that supports the rural development is the Chaipattana Foundation, which was initiated by the king himself and was officially registered in 1998. Chaipattana Foundation is funded from donations so it can accelerate development project implementations often restricted by bureaucratic procedures or budget constraints. It focuses on important development works that are restricted by regulations and are not overlapping with the government projects covering polluted water development, water capture to solve flood problems, environmental, and other community development projects such as occupational promotions, public health, and public welfare improvement.

With the objective to support the implementation of royally initiated and other development projects by collaborating with the government, charity organizations, and private society without any political involvement, the Chaipattana Foundation is deemed as a catalyst to the advancement of public wellbeing. The foundation also promotes development of social and economic welfare activities to enable them to become self reliant in the long run.

NOTES

1. http://www.chaipat.or.th/chaipat_old/noframe/eng/index.html, http://www.royalprojectthailand.com/general/english/main.html, http://kanchanapisek.or.th/index.en.html, http://www.adbi.org/conf-seminar-papers/2007/10/30/2390.csr.wedel/, http://goliath.ecnext.com/coms2/gi_0199-5739218/Interpretations-of-CSR-in-Thai.html, http://www.bangkokpost.com/60yrsthrone/working/index.html, http://thailand.prd.go.th/ebook_bak/story.php?idmag=31&idstory=238.
2. http://en.wikipedia.org/wiki/Thailand#Economy.
3. http://thailand.prd.go.th/view_inside.php?id=4574.
4. http://www.mfa.go.th/royalweb/index.html.

References

Clay, Jason. *Exploring the Links between International Business and Poverty Reduction: A Case Study of Unilever in Indonesia* (An Oxfam GB, Novib, Oxfam and Unilever Indonesia joint research project, Oxfam GB, Novib Oxfam, Netherlands, and Unilever, 2005).

Global Reporting Initiatives, Sustainability Reporting Guidelines, 2002.

Hollender, Jeffrey and Fenichell, Stephen. *What Matters Most—Business, Social Responsibility and the End of the Era of Greed* (Perseus Books, December 2003).

Ideas with impact, Harvard Business Review on Corporate Responsibility, Harvard Business School Publishing Corporation 2003.

Kemp, Melody. *Corporate Social Responsibility in Indonesia, Qiuixotic Dream or Confident Expectation?* (United nations Research Institute for Social Development, December 2001).

Kotler, Philip and Lee, Nancy. *Corporate Social Responsibility, Doing the Most Good for Your Company and Your Cause* (John Wiley & Sons, Inc.: Hoboken, New Jersey, 2005).

Mirvis, Philip, Ph.D and Googins, Bradley K., Ph.D.*Stages of Corporate Citizenship: A Developmental Framework* (The Center for Corporate Citizenship at Boston College: Massachusetts, 2006, http://www.bc.edu/corporatecitizenship).

National Center for Sustainability Reporting, http://www.ncsr-id.org/id/index.html.

Reputation: Risk of Risks, Economist Intelligence Unit Report, December 2005.

Rochlin, Steven A. and Googins, Bradley K., Ph.D.*The Value Proposition for Corporate Citizenship* (The Center for Corporate Citizenship at Boston College: Massachusetts, 2005, http://www.bc.edu/corporatecitizenship)

Statement on International Corporate Social Responsibility, Sociaal-Economische Raad, December 19, 2008, http://www.ser.nl/~/media/Files/Internet/Talen/Engels/2008/b27428/b27428_en.ashx.

"The Enterprise of the Future", IBM Global CEO Study 2008, http://www-935.ibm.com/services/us/gbs/bus/html/ceostudy2008.html.

WEBSITE LINKS FOR CASE STUDIES

Tata

Official website, http://www.tata.com
http://www.tata.com/ourcommitment/articles/inside.aspx?artid=cZrbJBDWkRQ=
http://www.tata.com/ourcommitment/articles/inside.aspx?artid=OifnYoPCmQ4=

PT Bank Danamon, Tbk

Corporate profile, http://www.danamon.co.id/content_a.php?nmCat=sekilas&lng=2
Danamon Peduli, http://www.danamonpeduli.or.id/en/about
Danamon Peduli Annual Report 2008, http://www.danamonpeduli.or.id/index.php?op=laporan&mode=detail&id=3&lang=en

PT Astra International, Tbk

Astra CSR Report 2006 and Astra Sustainability Report 2007, http://www.astra.co.id/sr.asp
Company profile, http://www.astra.co.id/article.asp?id=1000057
http://www.astra.co.id/article.asp?id=1000135
GCG & CSR, http://www.astra.co.id/article.asp?id=1000060
Yayasan Dharma Bhakti Astra (YDBA), http://www.ydba.astra.co.id/page.asp?id=4&pg=page&lv1=Profil&lv2=Sejarah%20YDBA&ver=ind
Astra Green Company, http://www.asria.org/ref/countries/lib/indonesia_Astra_Green_Company.pdf

PT Indo Tambangraya Megah, Tbk (ITM)

http://www.enggang.com
PT Indo Tambangraya Megah Tbk Annual Report 2008, http://www.itmg.co.id/en/investor/financial-information/annual-report

H.J. Heinz

Official website, http://www.heinz.com/
Heinz CSR Report 2007, http://www.heinz.com/CSR_2007/index.html
Heinz foundation, http://www.heinz.com/sustainability/social/heinz-foundation.aspx
Social sustainability, http://www.heinz.com/sustainability/social.aspx
Sustainability, http://www.heinz.com/sustainability.aspx

Related links

Micronutrient campaign, http://www.hki.org/programs/micronutrient.html
Heinz CSR news, http://www.csrwire.com/profile/9145.html

TNT International B. V., the Netherlands

Official website, http://www.tnt.com/
Official website (GCG, CSR reports), http://www.group.tnt.com/
Global Responsibility, http://www.tnt.com/express/en_id/site/home/about_us.html
2007 Sustainability Report, TNT NV, http://www.corporateregister.com
TNT annual report 2007 homepage, http://group.tnt.com/annualreports/annualreport07/index.html
TNT CSR, http://www.tnt.com/express/en_gb/site/home/about_us/about_tnt_express/tnt_corporate social_responsibility.html

Related links

TNT CSR case study, http://www.article13.com/A13_ContentList.asp?strAction=GetPublication&PNID=1367

Intel Corporation, USA

Official website, http://www.intel.com/
Intel CSR (Report, blog), http://www.intel.com/intel/corpresponsibility/index.htm?iid=gg_about+intel_gcr
CSR@Intel, http://blogs.intel.com/csr/
2007 CSR Report, http://www.intel.com/intel/cr/gcr/overview.htm

Motorola

Official website, http://www.motorola.com
http://www.motorola.com/responsibility

CSR in Thailand

http://en.wikipedia.org/wiki/Thailand#Economy
http://thailand.prd.go.th/view_inside.php?id=4574
http://www.chaipat.or.th/chaipat_old/noframe/eng/index.html
http://www.royalprojectthailand.com/general/english/main.html
http://kanchanapisek.or.th/index.en.html
http://www.adbi.org/conf-seminar-papers/2007/10/30/2390.csr.wedel/
http://goliath.ecnext.com/coms2/gi_0199-5739218/Interpretations-of-CSR-in-Thai.html
http://www.bangkokpost.com/60yrsthrone/working/index.html
http://thailand.prd.go.th/ebook_bak/story.php?idmag=31&idstory=238
http://www.mfa.go.th/royalweb/index.html

Index